A FARMER'S PRIMER ON GROWING RICE

Benito S. Vergara

1979
INTERNATIONAL RICE RESEARCH INSTITUTE
LOS BAÑOS, LAGUNA, PHILIPPINES
P.O. BOX 933, MANILA, PHILIPPINES

The International Rice Research Institute (IRRI) was established in 1960 by the Ford and Rockfeller Foundations with the help and approval of the Government of the Philippines. Today IRRI is one of the 13 nonprofit international research and training centers supported by the Consultative Group for International Agricultural Research (CGIAR). The CGIAR is sponsored by the Food and Agriculture Organization (FAO) of the United Nations, the International Bank for Reconstruction and Development (World Bank), and the United Nations Development Programme (UNDP). The CGIAR consists of 50 donor countries, international and regional organizations, and private foundations.

IRRI receives support, through the CGIAR, from a number of donors including: the Asian Development Bank, the European Economic Community, the Ford Foundation, the International Development Research Centre, the International Fund for Agricultural Development, the OPEC Special Fund, the Rockefeller Foundation, the United Nations Development Programme, the World Bank, and the international aid agencies of the following governments: Australia, Canada, China, Denmark, France, Federal Republic of Germany, India, Italy, Japan, Mexico, Netherlands, New Zealand, Norway, Philippines, Saudi Arabia, Spain, Sweden, Switzerland, United Kingdom, and United States.

The responsibility for this publication rests with the International Rice Research Institute.

Copyright © International Rice Research Institute 1986

All rights reserved. Except for quotations of short passages for the purpose of criticism and review, no part of this publication may be reproduced, stored in retrieval systems, or transmitted in any form or by any means, electronic, mechanical, photocopying, recording, or otherwise, without prior permission of IRRI. This permission will not be unreasonably withheld for use for noncommercial purposes. IRRI does not require payment for the noncommercial use of its published works, and hopes that this copyright declaration will not diminish the bona fide use of its research findings in agricultural research and development.

The designations employed and the presentation of the material in this publication do not imply the expression of any opinion whatsoever on the part of IRRI concerning the legal status of any country, territory, city, or area, or of its authorities, or concerning the delimitation of its frontiers or boundaries.

ISBN 971-104-051-4

Contents

1	Life cycle of the rice plant
9	The seed
19	Seedling growth
29	How to select good seedlings
37	Transplanting
43	The leaves
49	The roots
65	The tillers
77	The panicle
85	Dormancy
91	Fertilizers
99	How much nitrogen to apply
107	How to increase the efficiency of nitrogen fertilizer
117	Why more nitrogen fertilizer is applied during the dry season
123	Carbohydrates production
133	Water
141	Yield components
155	Plant type of a lowland rice variety with high grain potential
167	Factors affecting lodging
177	Weeds
189	Control of weeds
197	Herbicides
209	How to judge a rice crop at flowering

Foreword

A PROGRESSIVE RICE FARMER should understand *why* and *how* the improved rice varieties and farm technology increase production. But recommendations given to farmers often fail to answer questions such as why a farmer incubates seed, why he or she applies fertilizer, or how and when that fertilizer should be incorporated.

The farmer needs this knowledge to adjust his practices to suit his own unique farm situation.

To improve the understanding of rice culture among farmers, technicians, teachers, and scientists, B. S. Vergara of the IRRI plant physiology department prepared this handbook. *A farmer's primer on growing rice* should be particularly useful to technicians and the farmers that they serve. Dr. Vergara initiated work on the handbook during a sabbatical leave at the Southeast Asian Regional Center for Graduate Study and Research in Agriculture (SEARCA), Los Baños, Philippines. Donald Esslinger, editor of the Missouri State Agricultural Extension Service, took the leadership in editing *A farmer's primer on growing rice* while on sabbatic leave with IRRI's Office of Information Services (OIS).

N. C. Brady
Director General
International Rice Research Institute

LIFE CYCLE OF THE RICE PLANT

3 The rice plant
4 Growth stages of the rice plant
5 Difference in growth stages
6 Vegetative phase
7 Productive phase
8 Ripening phase

THE RICE PLANT

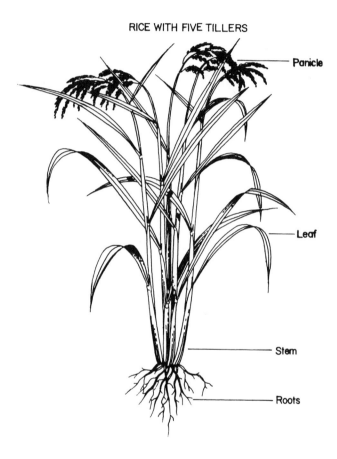

- A tiller is a shoot which includes the roots, stem, and leaves. It may or may not have a panicle.

GROWTH STAGES OF THE RICE PLANT

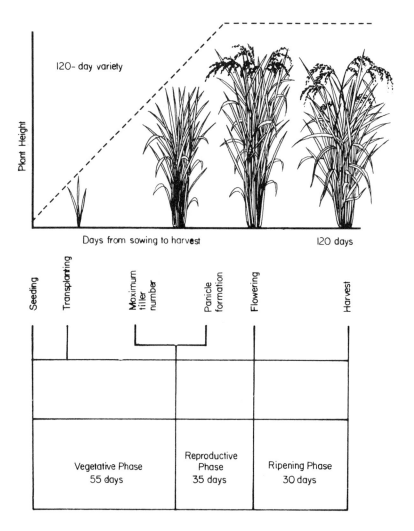

- Days in vegetative phase differ with variety.
- Reproductive and ripening phases are constant for most varieties. Panicle formation to flowering is 35 days. Flowering to harvest requires 30 days.
- Sowing to harvest may be 180 days or more.

DIFFERENCE IN GROWTH STAGES

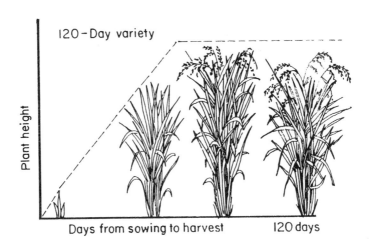

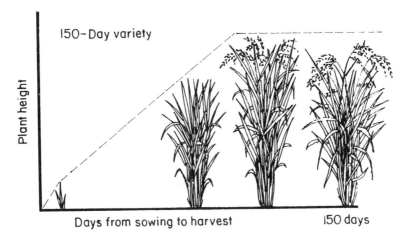

- Number of days in vegetative phase changes by variety.
- Days in reproductive and ripening phases are more or less fixed.
- Difference in growth duration is determined by days in vegetative phase.

VEGETATIVE PHASE

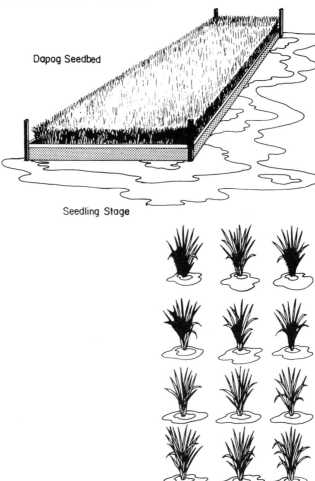

Dapog Seedbed

Seedling Stage

Tillering Stage

- Duration of the seedling or nursery stages:
  - Dapog (9–11 days)
  - Wetbed (16–20 days)
  - Direct seeding (none)
- Number of tillers and leaf area increase during the vegetative stage.
- Low temperature or long day length can increase the duration of the vegetative phase.

PRODUCTIVE PHASE

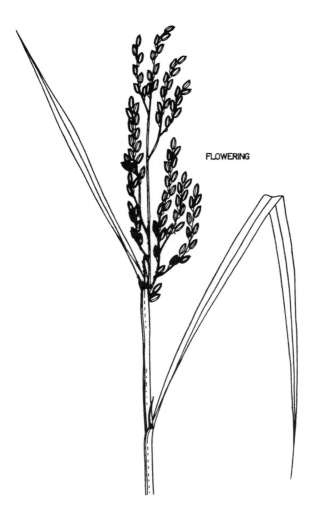

FLOWERING

- The reproductive phase begins at the start of panicle formation and ends at flowering. This takes around 35 days.

RIPENING PHASE

- The ripening phase starts at flowering and lasts for 30 days.
- Rainy days or low temperatures may delay the ripening phase, while sunny and warm days shorten it.
- To obtain high grain yields, good farming practices are needed at each growth stage.

THE SEED

11 The seed
12 Parts of a seed
13 Stages of germination
14 Conditions needed for seed germination — water
15 Conditions needed for seed germination — air
16 Conditions needed for seed germination — warm temperature
17 Why incubate the seeds
18 Why select good seeds

THE SEED

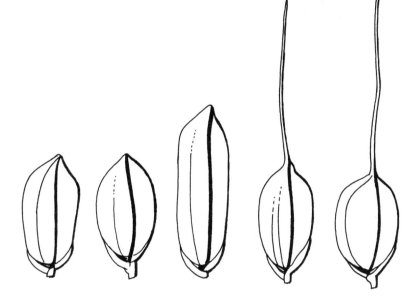

- Seeds vary in size, shape, color, and length of awn.

PARTS OF A SEED

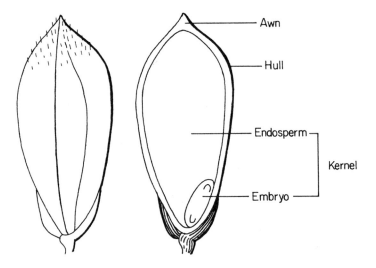

The seed was cut lengthwise

- The hull is the hard cover of the seed.
- The endosperm is made up mostly of starch, sugar, protein and fats. It is the storehouse of food for the embryo.
- Almost 80 percent of the endosperm is starch. The food needed for seed germination is in the endosperm.
- The embryo develops into the shoot and the roots.

STAGES OF GERMINATION

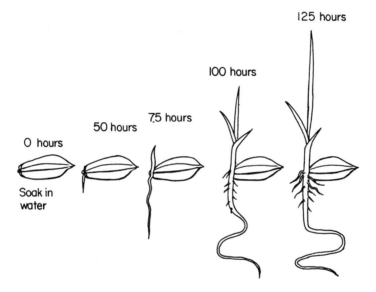

- Growth of the embryo will depend on temperature and availability of water and air.

CONDITIONS NEEDED FOR SEED GERMINATION – WATER

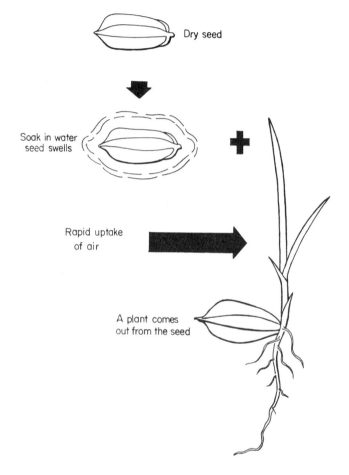

- Uptake of water is the first need of a seed for germination.
- There are many activities going on inside the germinating seed. Starch, protein and fats are being changed into simple foods for the embryo.
- Soak seeds for at least 24 hours so that water can easily and uniformly enter the seeds.

CONDITIONS NEEDED FOR SEED GERMINATION – AIR

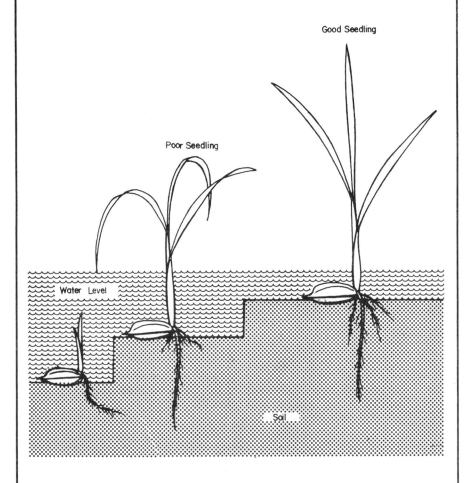

- The germinating rice seed needs air to live.
- Water contains very little air.
- If the seed is covered too deeply with water, the growth of the embryo will be slow and the resulting shoot is tall and weak.

CONDITIONS NEEDED FOR SEED GERMINATION—WARM TEMPERATURE

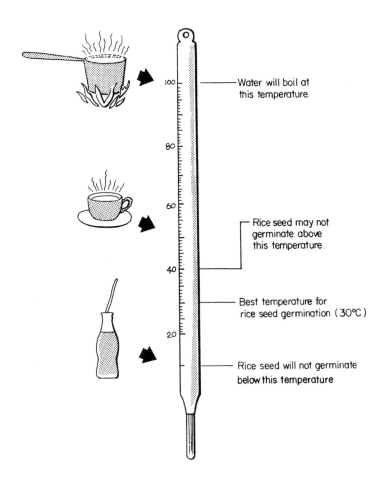

- Water will boil at this temperature.
- Rice seed may not germinate above this temperature.
- Best temperature for rice seed germination (30°C)
- Rice seed will not germinate below this temperature

- Warm temperature is needed to increase the activities inside the seed and thus increase growth.
- Low temperature decreases activities inside the seed.

WHY INCUBATE THE SEEDS

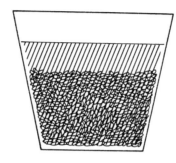

Soak 24 hours

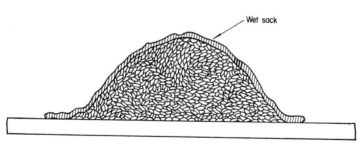

Incubate 24 hours

After soaking for 24 hours, water is removed and seeds are washed, placed on a cement floor and covered with a sack.

- Incubation keeps the seeds warm, increases growth of the embryo, and results in uniform germination.
- If the incubation temperature is too high, germination rate decreases and may actually kill the sprouted seeds.

WHY SELECT GOOD SEEDS

- The larger amount of food in good seeds results in better growth of seedlings.
- Good seeds result in better seedlings — healthier, heavier, and more roots.
- Healthy seedlings will grow faster than the poor seedlings when transplanted in the field.
- Good seeds result in uniform germination and growth.

SEEDLING GROWTH

21 Source of food for growth
22 Factors affecting seedling growth — water depth
23 Factors affecting seedling growth — amount of water
24 Factors affecting seedling growth — temperature
25 Factors affecting seedling growth — light intensity
26 Factors affecting seedling growth — light intensity
27 Factors affecting seedling growth — available nutrients
28 Factors affecting seedling growth — available nutrients

SOURCE OF FOOD FOR GROWTH

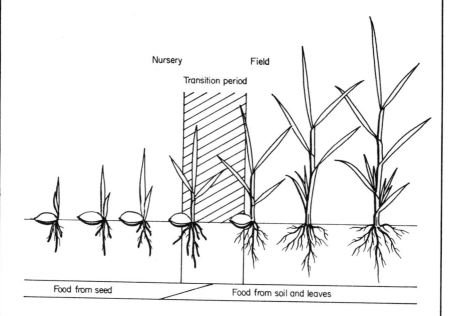

- The seedling grows by using food from the endosperm.
- After producing four leaves, it grows from food taken up through the roots and manufactured in the leaves.
- As the seedling gets older, it becomes more dependent on the environment for food.
- A dapog seedling contains very little food in the endosperm at the time of transplanting. It is just becoming independent in manufacturing its own food.

FACTORS AFFECTING SEEDLING GROWTH — WATER DEPTH

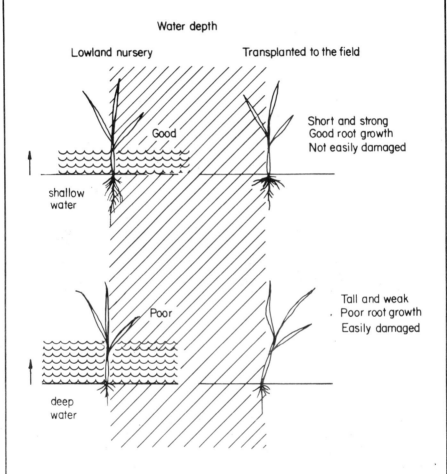

- Deep flooding results in poor root growth and tall seedlings due to lack of air in the soil. When transplanted, the seedlings are easily damaged.

FACTORS AFFECTING SEEDLING GROWTH— AMOUNT OF WATER

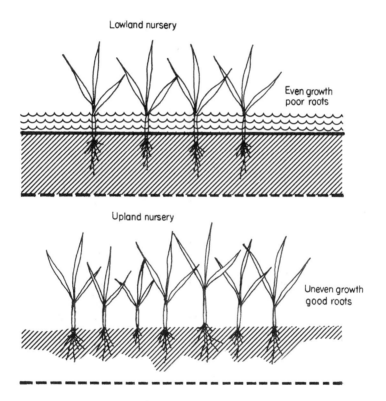

- The large amount of water available to the plants in lowland nursery results in uniform shoot growth.
- The irregular distribution of water in upland nursery results in uneven growth. However, root growth is usually excellent.
- Insufficient water results in slow seedling growth.

FACTORS AFFECTING SEEDLING GROWTH— TEMPERATURE

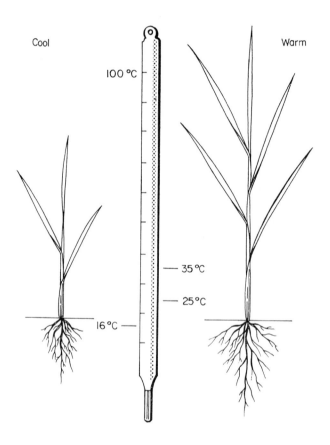

- Plants grow faster at warm temperature than at cooler temperatures.
- Seedlings are taller when grown during warm temperatures than during cooler temperatures. Cooler temperatures can cause yellowing of leaves and eventual death of seedlings.

FACTORS AFFECTING SEEDLING GROWTH — LIGHT INTENSITY

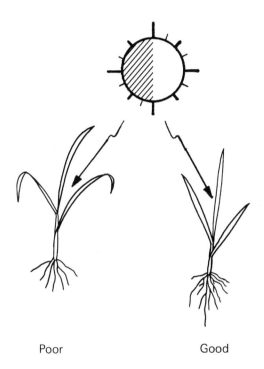

Poor   Good

- Seedlings need bright light. Cloudy days mean less brightness.
- Less light means weak seedlings because plants cannot produce enough food.
- Less light can cause elongation of leaf blade and leaf sheath — taller and weaker plants.
- Prepare seedbed away from shadows of trees and buildings.

FACTORS AFFECTING SEEDLING GROWTH — LIGHT INTENSITY

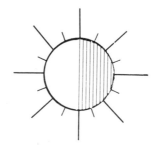

- Low light intensity results in

— tall and weak seedlings

— seedlings with low dry matter

— seedlings that are easily injured at planting

— increased chances of damage by diseases

FACTORS AFFECTING SEEDLING GROWTH — AVAILABLE NUTRIENTS

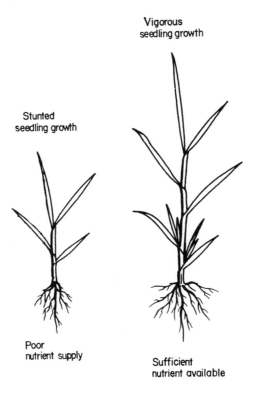

- Fertilizers supply plant food in addition to what is already available in the soil.
- Fertilizer may be needed if the nursery period is long, in upland nurseries, in cold areas, and in areas with poor soil.

FACTORS AFFECTING SEEDLING GROWTH — AVAILABLE NUTRIENTS

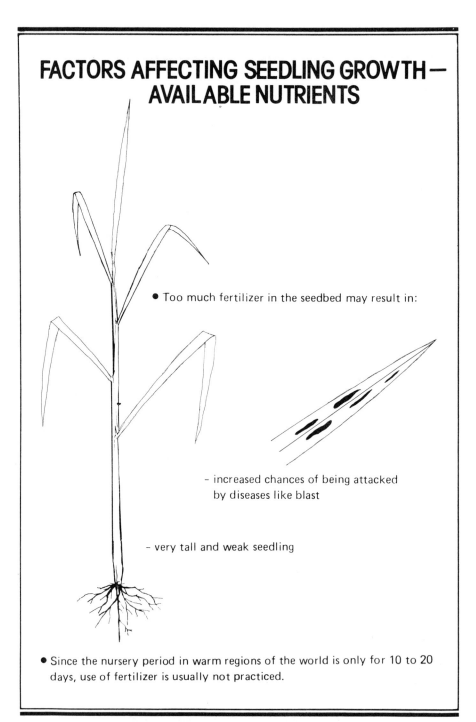

- Too much fertilizer in the seedbed may result in:

  - increased chances of being attacked by diseases like blast

  - very tall and weak seedling

- Since the nursery period in warm regions of the world is only for 10 to 20 days, use of fertilizer is usually not practiced.

HOW TO SELECT GOOD SEEDLINGS

31 Good seedlings have uniform plant height and growth
32 Good seedlings have short leaf sheath
33 To have short leaf sheaths use proper water depth
34 To have short leaf sheaths good lighting is needed
35 Good seedlings have no pests nor diseases such as
36 Good seedlings have large number of roots and heavy weight

GOOD SEEDLINGS HAVE UNIFORM PLANT HEIGHT AND GROWTH

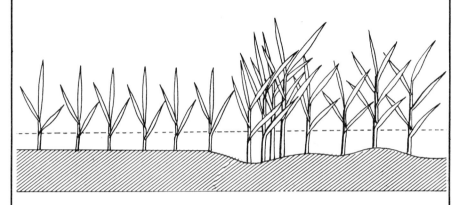

Regular plant growth   Irregular plant growth

- Irregular seedling growth may indicate unevenness in:
  - distribution of seeds on the seedbed
  - germination
  - land preparation of the seedbed
  - watering
  - availability of nutrients in the soil

GOOD SEEDLINGS HAVE SHORT LEAF SHEATH

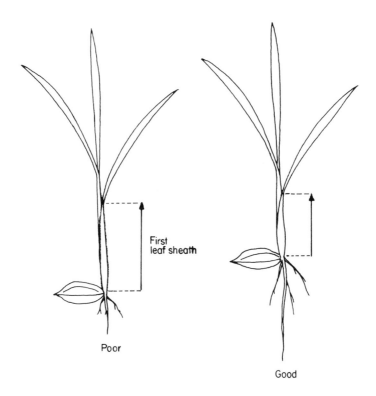

- The leaf sheath is the lower portion of the leaf that encloses the stem and young leaves.
- Long leaf sheath indicates very rapid initial elongation, making the seedling weak.

TO HAVE SHORT LEAF SHEATHS USE PROPER WATER DEPTH

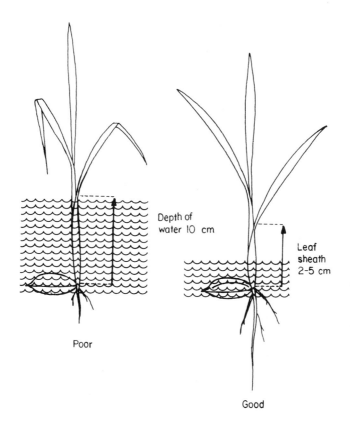

Water level on the seedbed

- Too much water can result in long leaf sheaths and weak seedlings.
- Weak seedlings have poor growth right after transplanting, recovery is slow.
- Long, droopy leaves of poor seedlings tend to stick to the mud when transplanted.

TO HAVE SHORT LEAF SHEATHS GOOD LIGHTING IS NEEDED

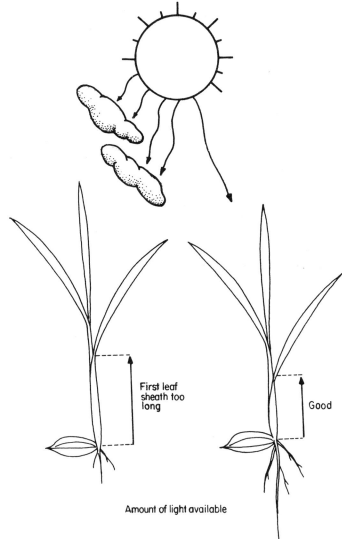

- Low light because of cloudy days, heavy seeding, and shadows from trees can result in long leaf sheaths.

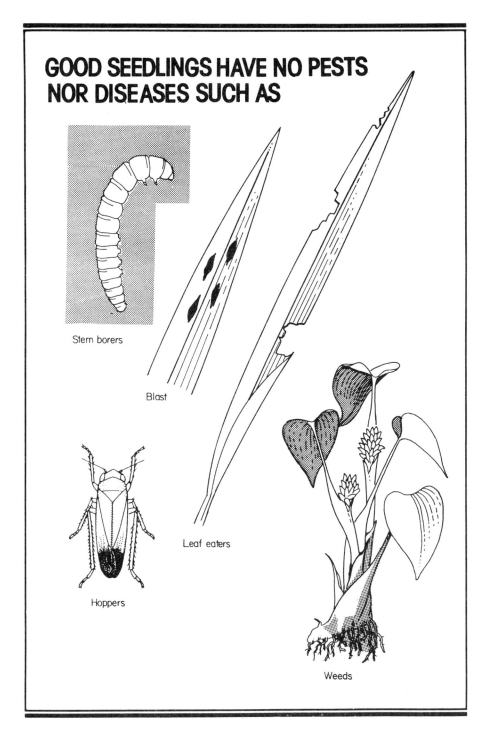

GOOD SEEDLINGS HAVE LARGE NUMBER OF ROOTS AND HEAVY WEIGHT

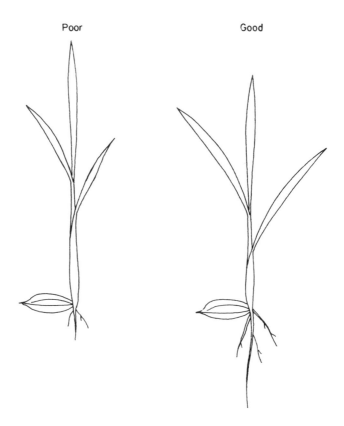

- Heavy seedlings indicate large amount of food accumulated, resulting in better recovery after transplanting.

TRANSPLANTING

39 Why transplant
40 How many seedlings per hill
41 Why transplant at the proper depth
42 Why cut the leaves of seedlings before transplanting

WHY TRANSPLANT

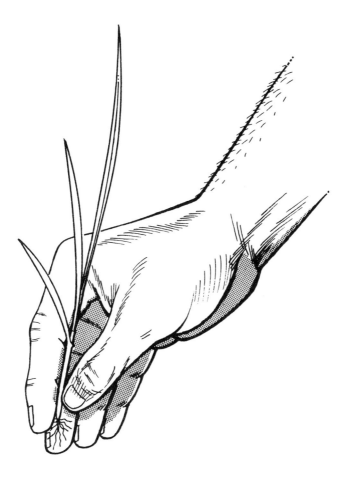

- Weed control is simpler in straight row transplanting.
- Direct-seeded rice is susceptible to attack by rats, snails and birds.

HOW MANY SEEDLINGS PER HILL

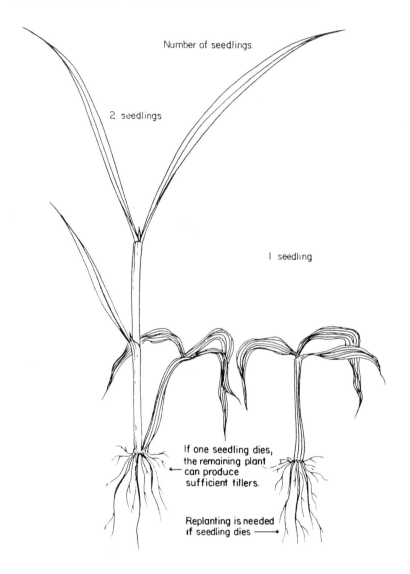

- There is no difference in grain yield between one and two seedlings per hill, if no seedlings die.

WHY TRANSPLANT AT THE PROPER DEPTH

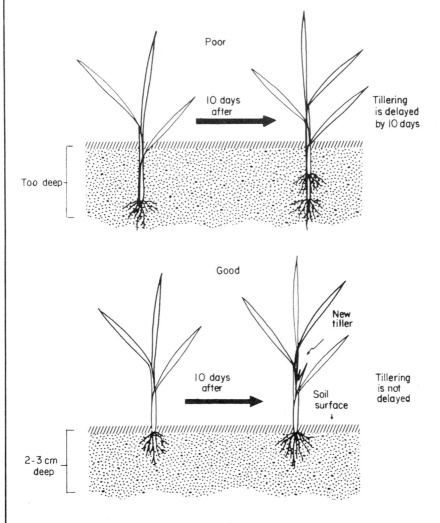

- Tillers normally develop 5 to 10 days after transplanting. Deep planting delays tillering.
- Growth of the plant is set back at transplanting; it takes 2 to 4 days before new roots are formed.

WHY CUT THE LEAVES OF SEEDLINGS

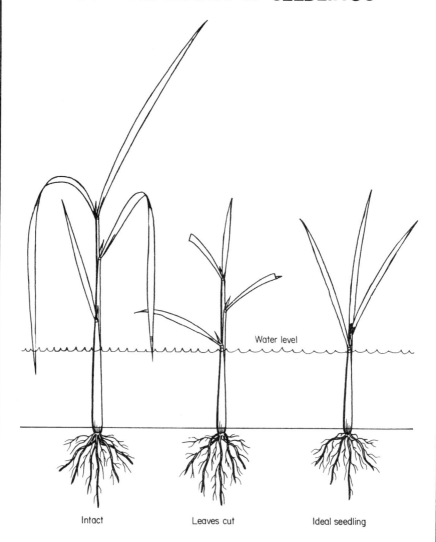

- Long, droopy leaves of tall seedlings touch the muddy water providing chance of diseases infecting the leaves. Cutting prevents this.
- Wounds caused by cutting may serve as entrance for bacterial infection. To avoid cutting, plant seedlings of right age grown under good conditions.

THE LEAVES

45 The rice leaf
46 The leaves of the main stem
47 Leaf production
48 Internodes

THE RICE LEAF

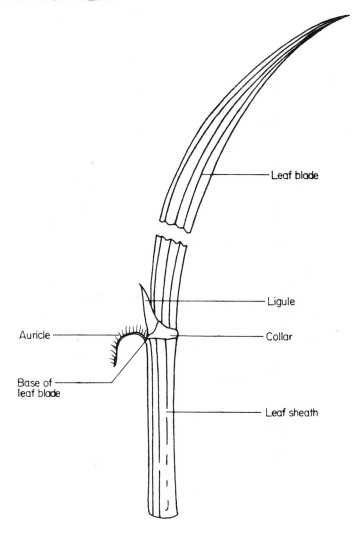

- A rice leaf can be distinguished from other grasses by the presence of both the ligule and the auricle.
- A grass leaf has a collar but may have only a ligule or auricle or neither.
- A rice leaf, like the grasses, has parallel veins.

THE LEAVES

THE LEAVES OF THE MAIN STEM

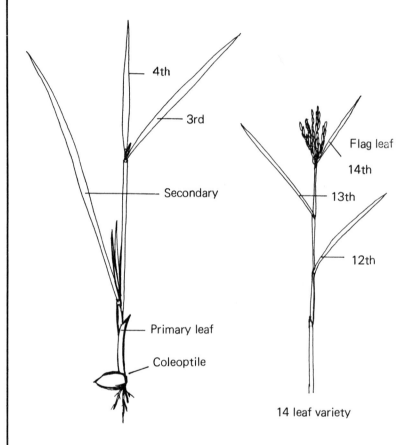

- The coleoptile comes out of the seed first, then the primary leaf, the secondary leaf with the first expanded leaf blade, and succeeding leaves.
- The last leaf is called the flag leaf.

LEAF PRODUCTION

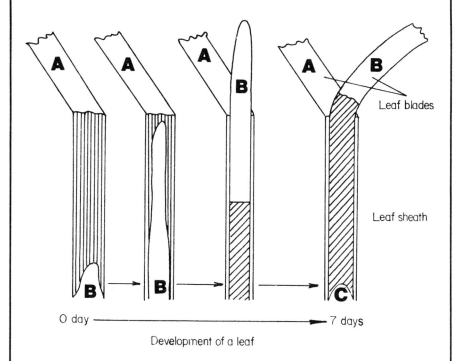

LENGTHWISE SECTION AT THE TIP OF THE STEM

Development of a leaf

- Rice leaves on the main stem are produced one at a time.
- A new leaf is produced at an average of every 7 days.
- Rice leaves are alternately arranged.

THE LEAVES

INTERNODES

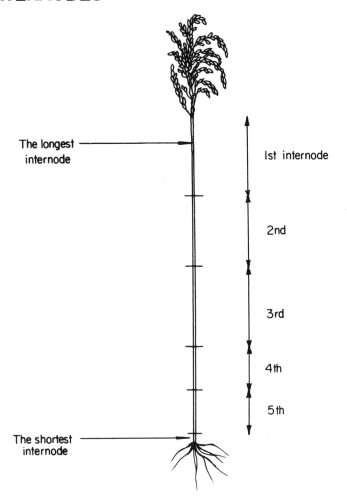

- Normally there are four to six elongated (more than 1 cm) internodes at harvest.
- The longer the basal internodes, the greater the tendency for the plant to fall flat on the ground.
- Closer planting, cloudy weather, higher nitrogen level of the soil and higher temperatures will result in longer internodes.

THE ROOTS

51	Origin of roots
52	Crown roots
53	Root hairs
54	Functions of the roots — site of water and nutrient uptake supports the upper parts of the plant
55	Root development
56	Root development — 30 days after transplanting
57	Root development — 50 days after transplanting
58	Root development at heading
59	Root distribution
60	Root distribution depends upon the depth of top soil
61	Root distribution depends upon the depth of the plowed layer
62	Root distribution depends upon the downward movement of water
63	Root distribution depends upon the amount of air available
64	Root distribution depends upon placement of fertilizer

ORIGIN OF ROOTS

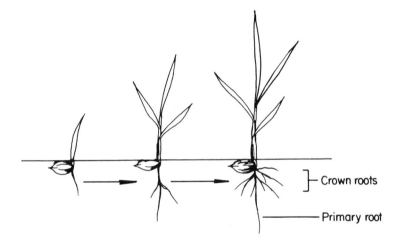

- The radicle or primary root usually dies within a month.
- Crown roots develop from the lower nodes.
- Old roots or older parts of a root are colored brown while new and young parts of a root are white.

CROWN ROOTS

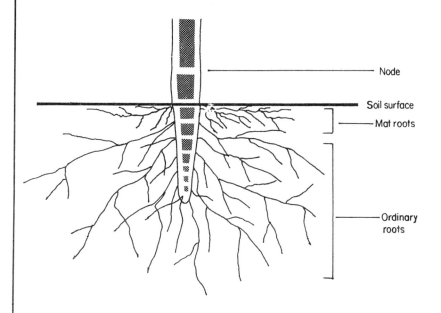

- Crown roots develop from the lower nodes.
- There are two types of crown roots, superficial and ordinary roots.
- Superficial roots develop easily when the air content of the soil is low, as in later growth stage.

ROOT HAIRS

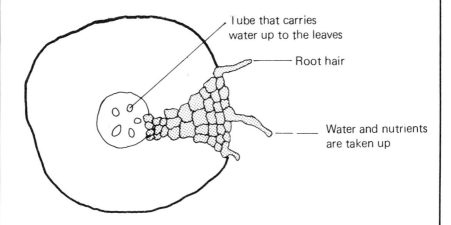

Cross section of a young root enlarged 120 times

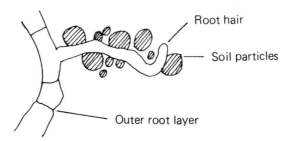

- Root hairs are tubular extensions on the outermost layer of the roots.
- They are important in water as well as nutrient uptake.
- Root hairs are generally short-lived.

FUNCTIONS OF THE ROOTS –
SITE OF WATER AND NUTRIENT UPTAKE
SUPPORTS THE UPPER PARTS OF
THE PLANT

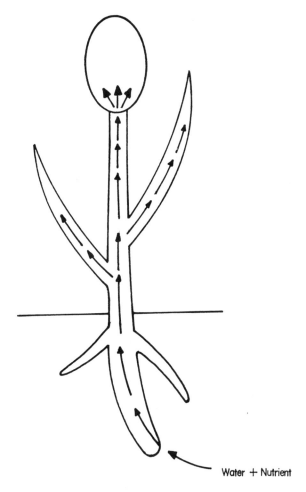

- Soil water contains nutrients such as nitrogen, phosphorus, and potassium.

ROOT DEVELOPMENT

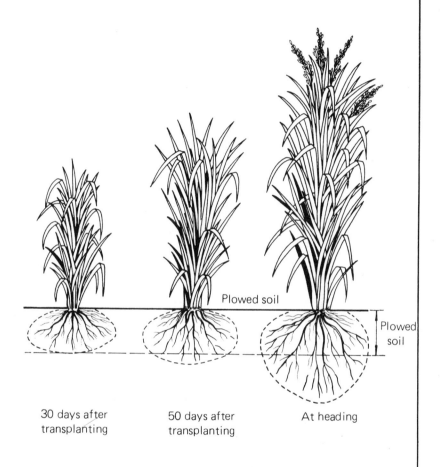

30 days after transplanting

50 days after transplanting

At heading

- At later stages of plant growth, the roots initiated from the upper nodes develop horizontally and form the superficial or mat roots.

ROOT DEVELOPMENT – 30 DAYS AFTER TRANSPLANTING

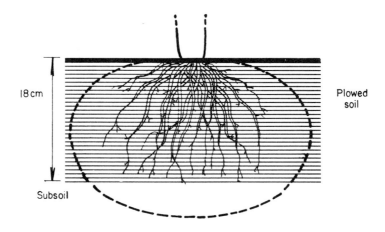

- The roots are found mostly within the plowed layer (18 cm), almost no roots in the subsoil.

ROOT DEVELOPMENT – 50 DAYS AFTER TRANSPLANTING

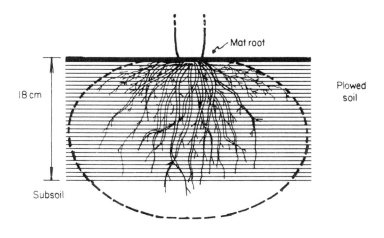

- Some roots have grown downward to the subsoil.

ROOT DEVELOPMENT AT HEADING

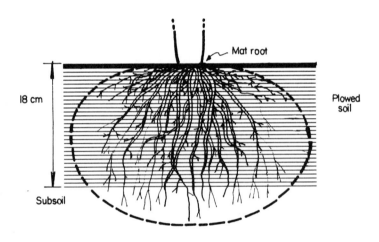

- Some big and strong roots have further penetrated into the subsoil.
- Mat roots are abundant.

ROOT DISTRIBUTION

Water

Fertilizer

Plowed soil

Hardpan

- Root distribution depends upon:
  - depth of the top soil
  - depth of the plowed layer
  - downward movement of water
  - amount of air available
  - type of irrigation
  - placement of fertilizer
- Roots must penetrate deeply and spread widely and evenly for good uptake of nutrient from the soil.

ROOT DISTRIBUTION DEPENDS UPON THE DEPTH OF TOP SOIL

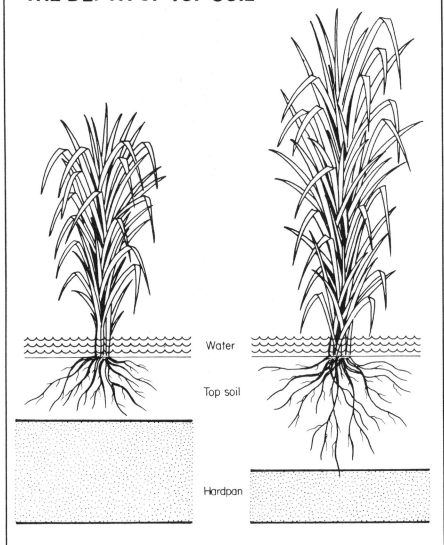

- The deeper the top soil or the distance between the surface and the hardpan, the deeper is the root penetration.

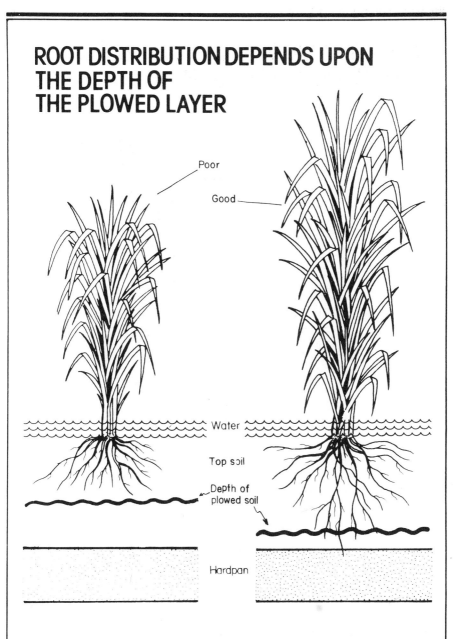

ROOT DISTRIBUTION DEPENDS UPON THE DEPTH OF THE PLOWED LAYER

- The deeper the plowed layer, the deeper the root penetration.
- Plow as deep as possible, shallow plowing restricts root growth.

ROOT DISTRIBUTION DEPENDS UPON THE DOWNWARD MOVEMENT OF WATER

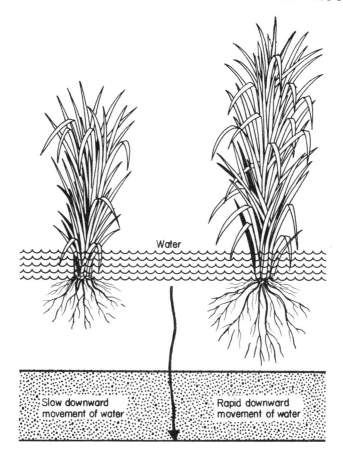

- If the downward movement of water happens freely and quickly, the roots develop downward easily.
- The downward movement or percolation of water results in more air and fertilizer available in the lower soil layer.
- The deeper the roots the better will be the water absorbing capacity of the plant. A very important plant characteristic in areas where water supply is not dependable.

ROOT DISTRIBUTION DEPENDS UPON THE AMOUNT OF AIR AVAILABLE

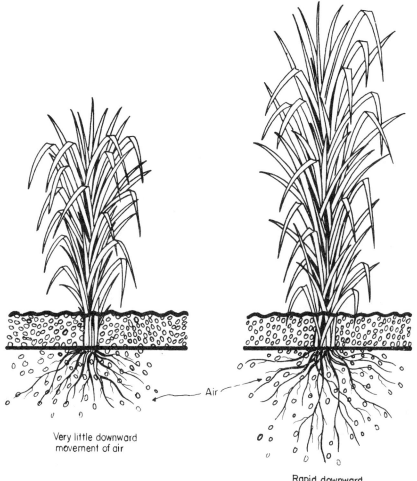

Very little downward movement of air

Rapid downward movement of air

- Absence of air at a soil layer can result in the decay of roots and inhibition of root development in that layer. A shallow root type develops.
- Downward movement of air dissolved in water depends on the depth and type of top soil.

ROOT DISTRIBUTION DEPENDS UPON PLACEMENT OF FERTILIZER

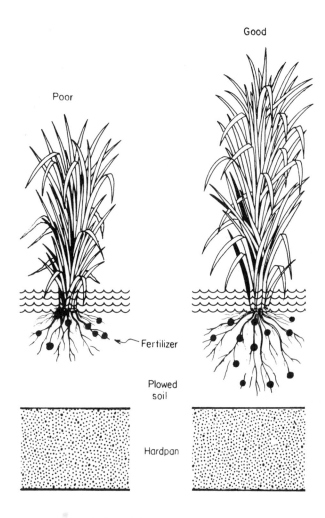

- Mixing the fertilizer thoroughly into the plowed soil results in deeper roots and better root distribution.

THE TILLERS

67 Primary tiller
68 Tillering pattern
69 Production of tillers
70 Productive and non-productive tillers
71 Percent productive tillers
72 Factors affecting tillering — variety
73 Factors affecting tillering — spacing
74 Factors affecting tillering — season of planting
75 Factors affecting tillering — nitrogen level

PRIMARY TILLER

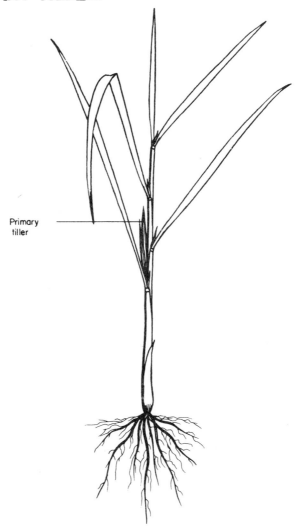

Primary tiller

- The first tiller (primary tiller) usually develops between the main stem and the second leaf from the base.
- Although the tiller is still attached to the mother plant at later stages of growth, it is independent since it produces its own roots.

TILLERING PATTERN

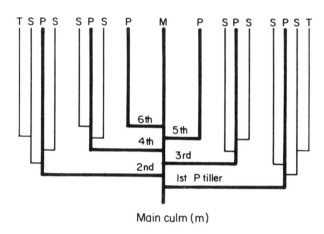

- Primary (P) tillers come from the main stem.
- Secondary (S) tillers develop from the primary tillers and tertiary (T) tillers from the secondary tillers.
- The lower the point of origin on the main stem, the older is the tiller.

PRODUCTION OF TILLERS

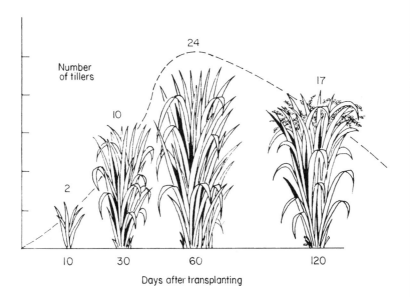

- Tillering starts 10 days after transplanting and reaches maximum 50 to 60 days after transplanting.
- After reaching the maximum, tiller number decreases as weak tillers die.

PRODUCTIVE AND
NON-PRODUCTIVE TILLERS

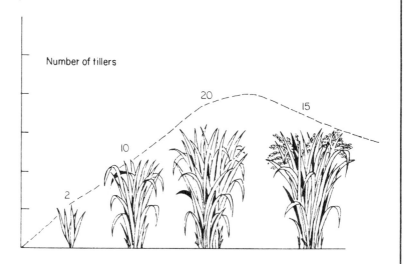

- Tillers formed at later stages of growth are usually unproductive. Either the tillers die or the panicles produced are small and too late to catch up with the ripening of the other panicles. The spikelets are only half-filled at the time of harvest.
- Modern varieties have more tillers at flowering and lose less tillers.
- Loss of tillers may be the result of mutual shading, competition among tillers, or lack of nutrients, especially nitrogen.

PERCENT PRODUCTIVE TILLERS

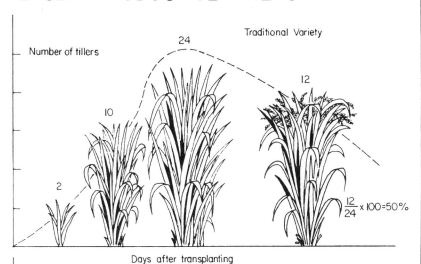

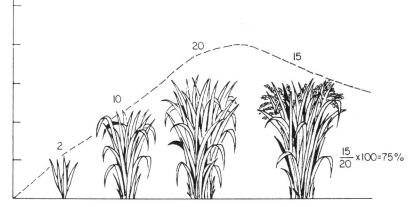

$$\text{Percent productive tillers} = \frac{\text{number of panicles produced}}{\text{highest number of tillers produced}} \times 100$$

In the above drawings, the percent productive tillers is 50 in the traditional varieties and 75 in the modern varieties.

THE TILLERS

FACTORS AFFECTING TILLERING— VARIETY

A variety with 19 tillers

A variety with 54 tillers

- Varieties differ in tillering capacity.
- Tillering capacity can be obtained by growing the plants far apart in rich soil. The full capacity usually is not used under field conditions.

FACTORS AFFECTING TILLERING — SPACING

50 cm

50 x 50 cm spacing
33 tillers per plant
4 plants per square meter
122 tillers per square meter

10 cm

10 x 10 cm spacing
3 tillers per plant
100 plants per square meter
300 tillers per square meter

- The tiller number per plant increases with increase in distance between plants.
- The number of tillers per square meter may be less if plants are spaced too far apart.

FACTORS AFFECTING TILLERING — SEASON OF PLANTING

Rainy season – 21 tillers

Dry season – 16 tillers

- More tillers are produced during the rainy than the dry season.
- More nitrogen fertilizer is needed during dry season to increase tiller number.

FACTORS AFFECTING TILLERING — NITROGEN LEVEL

10 Tillers
No nitrogen
added

30 Tillers
Fertilizer nitrogen
added

- The higher the amount of nitrogen added the more tillers produced.

THE PANICLE

79 Panicle formation
80 Booting stage
81 The spikelet
82 Flowering order of a panicle
83 Stages of grain formation
84 Causes of empty spikelets

PANICLE FORMATION

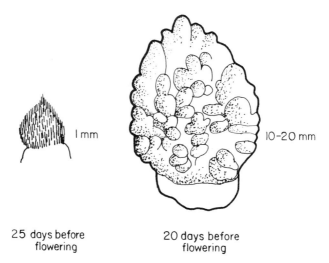

25 days before flowering

20 days before flowering

- Panicle formation occurs at the tip of the growing point of the shoot.
- The panicle is visible to the naked eye when it is 1 millimeter in length.
- At 1 millimeter, the young panicle has many fine, white, hairy structures at the tip.
- When the panicle inside the leaf sheath is about 1 mm, three more leaves will be produced before the panicle comes out.

BOOTING STAGE

- At booting stage, there is a bulge at the base of the leaf sheath.
- Booting stage is 20 to 25 days before flowering, the panicle is 1 millimeter in size.
- Flowering occurs 35 days after the start of panicle formation.

THE SPIKELET

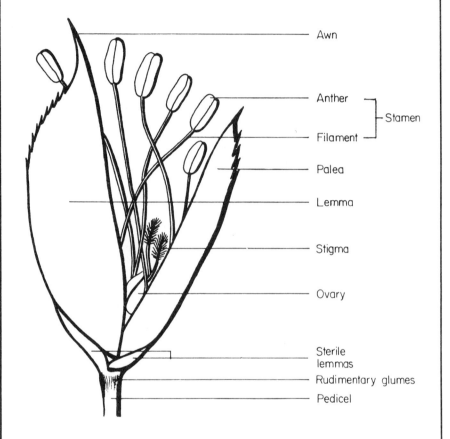

- Opening of the anthers occurs one day after the panicle comes out.
- Low temperature delays the opening of the anthers.
- Pollen from the anthers (like fine dust) have to reach the stigma and unite with the egg inside the ovary for the ovary to develop into a grain.

FLOWERING ORDER OF A PANICLE

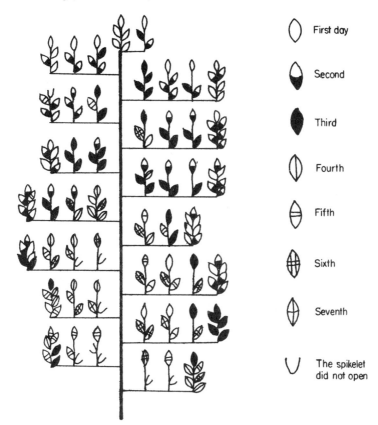

A large panicle with 196 spikelets

- First day
- Second
- Third
- Fourth
- Fifth
- Sixth
- Seventh
- The spikelet did not open

- Spikelets on the top branches open first.
- The lower spikelets, which open last, are usually not completely filled in some large panicles.
- Modern varieties have 100 to 120 spikelets per panicle.

STAGES OF GRAIN FORMATION

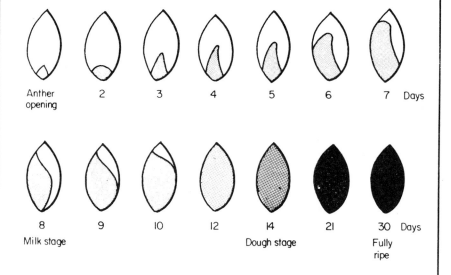

- Build-up of starch inside the spikelet begins after the sperm in the pollen unites with the egg in the ovary (fertilization).
- A grain is a ripened ovary together with the lemma and palea.
- At 21 days after fertilization the spikelet has reached its maximum weight.
- Since it takes 7 days for all the spikelets in a panicle to open, full maturity for the whole panicle does not occur until 10 days after flowering.
- Extra days will be needed to ripen all the grains since the panicles do not come out at the same time.

CAUSES OF EMPTY SPIKELETS

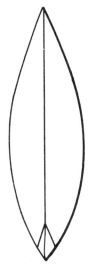

Side view of a fully filled spikelet

Side view of an empty spikelet

- Many factors can affect the filling up of the spikelets such as:
  - lack of starch to fill up the spikelets because of lodging, low light intensity, drying of the leaves, and diseases.
  - drying up of the stigma because of high temperature or dry winds.
  - too much nitrogen applied at panicle formation stage.
  - low temperature and high humidity at flowering resulting in the non-opening of the spikelets.
  - low temperature at panicle formation resulting in degeneration of the pollen.
- Empty spikelets will float when placed in water.

DORMANCY

87 Grain dormancy
88 Advantages of dormancy — prevents germination of seeds in the panicle
89 Advantages of dormancy — prevents germination of seeds if stored in wet conditions for a few days after harvest

GRAIN DORMANCY

Days after harvest				
0	(seed)	+ Water	→	Not dormant
0	(seed)	+ Water	→	Dormant
7	(seed)	+ Water	→	Still dormant
21	(seed)	+ Water	→	Still dormant
28	(seed)	+ Water	→	Not dormant

- Dormancy is the failure of a mature seed to germinate under favorable conditions.
- Not all varieties have dormancy.
- Seeds may be dormant from 0 to 80 days depending upon the variety and conditions at harvest.

DORMANCY

ADVANTAGES OF DORMANCY — PREVENTS GERMINATION OF SEEDS IN THE PANICLE

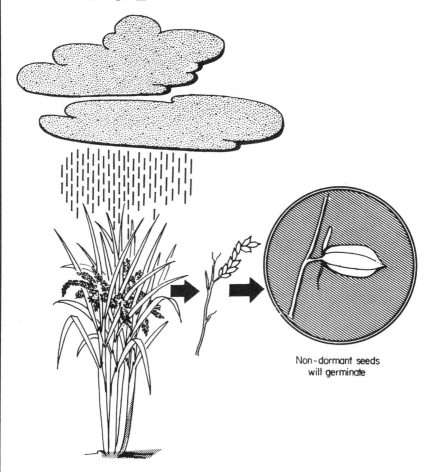

Non-dormant seeds will germinate

- Dormancy is important during the rainy season harvest.
- When the grains are mature and it rains, the nondormant seeds may germinate.
- Seeds harvested during the dry, sunny season have lower percentage of dormancy.

- The cause of dormancy is not clear.
- Dormancy can be a disadvantage, freshly harvested seeds cannot be immediately planted.

FERTILIZERS

93 What is a fertilizer
94 Nutrients that the rice plant needs
95 Role of fertilizers
96 Types of fertilizers — organic
97 Types of fertilizers — inorganic
98 Fate of nitrogen fertilizer applied to the soil

WHAT IS A FERTILIZER

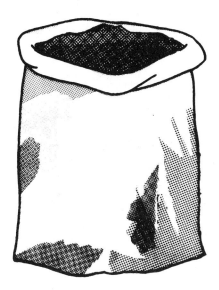

- Fertilizers contain important mineral nutrients needed by the plant and are usually applied to the soil.
- Soils sometimes do not provide sufficient nutrients that rice plants need.
- Fertilizers should be applied if mineral nutrients are lacking.

NUTRIENTS THAT THE RICE PLANT NEEDS

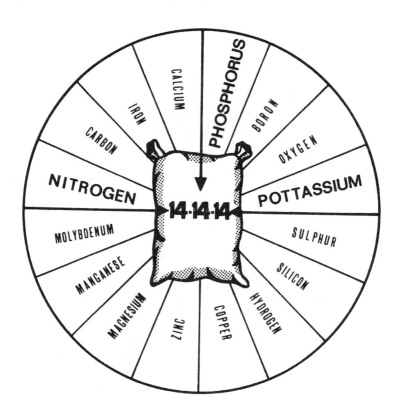

- All of the above except carbon, oxygen, and hydrogen can be supplied by fertilizers.
- There are several mineral nutrients that the plant needs but nitrogen, potassium, and phosphorus are needed in large amounts.

ROLE OF FERTILIZERS

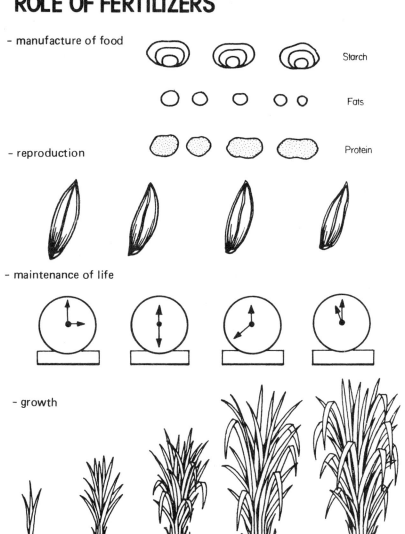

- manufacture of food — Starch, Fats, Protein
- reproduction
- maintenance of life
- growth

- Nitrogen, phosphorus, and potassium are needed for the life processes going on in the plant:

TYPES OF FERTILIZERS — ORGANIC

Examples:
Manure
Compost

Mineral Nutrient

Non-mineral Nutrient Material

Manure

- Organic fertilizers come from plant and animal matters such as rotten leaves and chicken manure.
- Large amounts of organic fertilizer contain very small amount of mineral nutrients needed by the plant.
- Use of organic fertilizer results in better soil structure.

TYPES OF FERTILIZERS— INORGANIC

Examples:

Urea (45-0-0)
Ammonium sulfate (21-0-0)
Muriate of potash (0-0-60)

- Inorganic fertilizers are commercially manufactured mineral nutrients.
- There are several kinds of combinations of nitrogen, phosphorus, and potassium fertilizers.
- The numbers on the bag refer to the percentage by weight of mineral nutrients in the fertilizer. 24—12—12 means 24% nitrogen, 12% phosphorus (P_2O_5), and 12% potassium (K_2O).
- The rest of the material in the fertilizer bag is filler material and may contain calcium or sulfur.

FATE OF NITROGEN FERTILIZER APPLIED TO THE SOIL

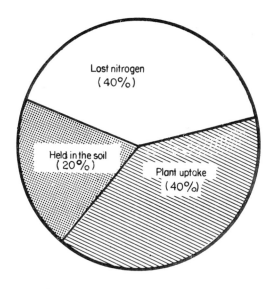

- A high percentage of the nitrogen applied is lost.
- Nitrogen fixed in the soil can be partly used by the succeeding crop.
- How to minimize the loss and maximize the use of the available nitrogen is important in good crop management.

HOW MUCH NITROGEN TO APPLY

101 Season of cropping — rainy season
102 Season of cropping — dry season
103 Fertility of the soil
104 Yield potential of the variety
105 Profit from fertilizer applied

SEASON OF CROPPING — RAINY SEASON

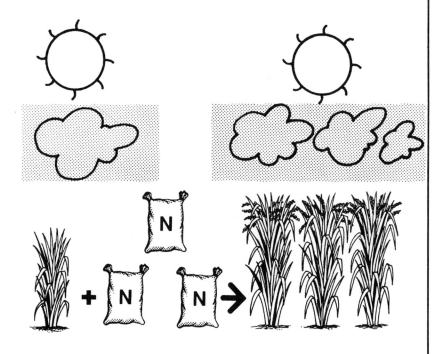

Dry season

- Rainy season — plants are tall, leafy, and shade each other so that food manufacturing in the leaves is low. Light energy is low above and inside the crop.
- Fertilizer applied during rainy season cannot be fully used by the plant.
- Lower amount of fertilizer should be used during the rainy season.

SEASON OF CROPPING — DRY SEASON

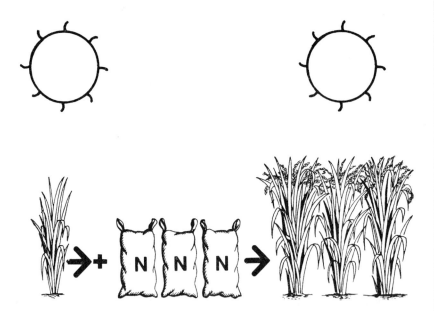

- Dry season — plants are shorter and have less tillers, light energy available is more.
- The fertilizer applied increases tiller number, leaf area, and rate of food manufacturing.
- More sunlight and more leaves increase food production — higher profits for fertilizer applied.
- More nitrogen can be applied during the dry season since grain yield as a result of nitrogen application is higher.

FERTILITY OF THE SOIL

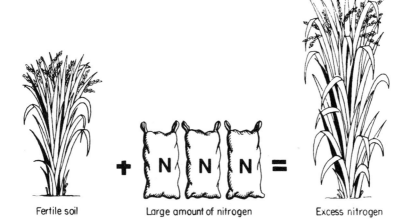

Fertile soil + Large amount of nitrogen = Excess nitrogen

- Too much nitrogen fertilizer in the soil results in too much vegetative growth, resulting in poor distribution of light and possibly lodging of plants.
- Too much nitrogen at later stages of growth increases sterility of spikelets and induces production of late tillers.

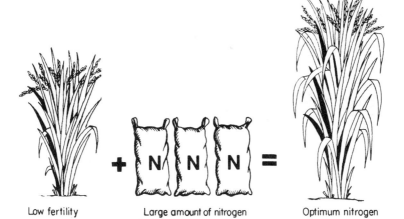

Low fertility + Large amount of nitrogen = Optimum nitrogen

- An optimum level of nitrogen in the soil results in the correct leaf area, tiller number and distribution of light, resulting in higher grain yields than the above soil.

YIELD POTENTIAL OF THE VARIETY

High yield potential — semidwarf

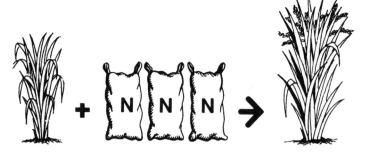

Nitrogen fertilizer

Upright leaves
increase tillers
no lodging

Low yield potential — tall

Nitrogen fertilizer

Droopy leaves
mutual shading
lodging

- Application of fertilizer to tall varieties will increase their height and tendency to lodge.
- Because of lodging and shading of leaves, grain yields may actually decrease with fertilizer application.

PROFIT FROM FERTILIZER APPLIED

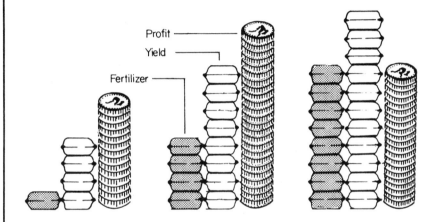

Low fertilizer	Medium fertilizer	High fertilizer
Low yield	Medium yield	Highest yield
Low profit	High profit	Low profit

- There is a right amount of fertilizer to apply to get maximum profit.
- The right amount of fertilizer will depend on the price of fertilizer in relation to increase in yields.
- The profit from fertilizer applied is higher during the dry than in the rainy season.
- The right amount of fertilizer for high grain yields will vary with the variety.

HOW TO INCREASE THE EFFICIENCY OF NITROGEN FERTILIZER

109	Use high yielding varieties
110	Apply the right amount of fertilizer
111	Apply fertilizer at correct growth stage of the rice plant
112	Prevent the field from drying out
113	Mix the fertilizer into the soil
114	Do not topdress while the leaves are wet
115	Keep the fields free from weeds

USE HIGH YIELDING VARIETIES

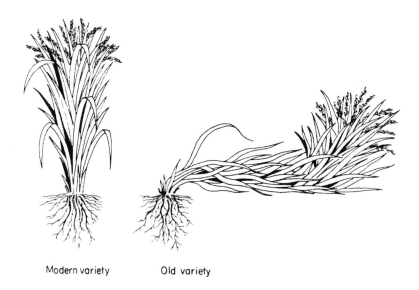

Modern variety   Old variety

Comparative grain yields

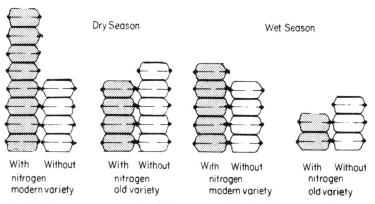

Dry Season		Wet Season	
With Without nitrogen modern variety	With Without nitrogen old variety	With Without nitrogen modern variety	With Without nitrogen old variety

- Grain yield increase as a result of nitrogen application is more in the modern than in the old varieties, regardless of the season of planting or amount of nitrogen used.

APPLY THE RIGHT AMOUNT OF FERTILIZER

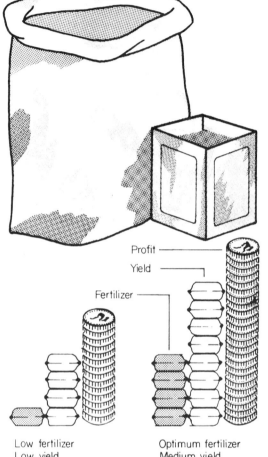

Low fertilizer
Low yield
Low profit

Optimum fertilizer
Medium yield
High profit

High fertilizer
Highest yield
Low profit

- The right amount of fertilizer will depend upon:
  - season of cropping
  - fertility of the soil
  - yield potential of the variety
  - price of the fertilizer
  - time and method of application

APPLY FERTILIZER AT CORRECT GROWTH STAGE OF THE RICE PLANT

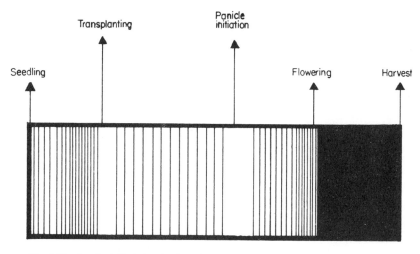

The lighter the shade the better the time of application

- The most efficient times for applying nitrogen fertilizers are at transplanting and at panicle initiation.

PREVENT THE FIELD FROM DRYNG OUT

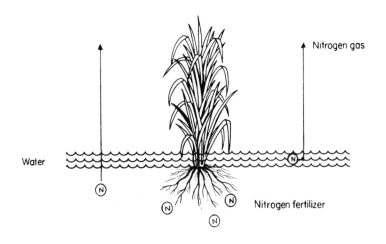

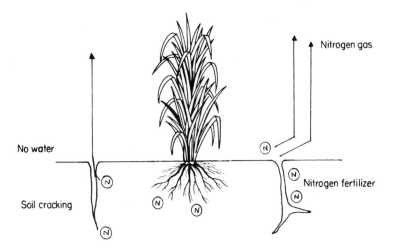

- On and off flooding results in great losses of nitrogen to the air.
- Nitrogen applied to flooded soil is changed to a different form by the air. This form is easily changed into gas and lost into the air.
- Water keeps the air from moving into the soil. The less air in the soil the less change in nitrogen to gas form. Repair levees to prevent water loss.

MIX THE FERTILIZER INTO THE SOIL

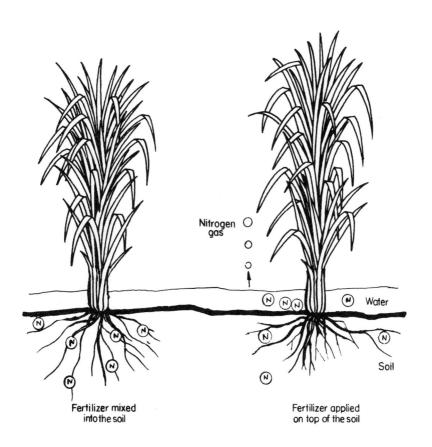

- Fertilizers applied before transplanting should be mixed thoroughly with the soil:
   - to prevent nitrogen losses to the atmosphere by the action of air.
   - to put the fertilizer nearer to the roots.
- Do not broadcast fertilizer without mixing it into the soil.
- Do not topdress in water immediately after transplanting.

DO NOT TOPDRESS
WHILE THE LEAVES ARE WET

- The fertilizer will stick on the wet leaves and may cause leaf burn.
- The dissolved fertilizer will be lost to the air when the droplets dry up.

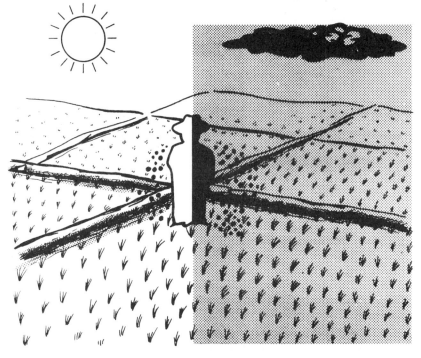

- Similarly do not topdress if a heavy rain is impending: The fertilizer might be washed out from the field.

KEEP THE FIELDS FREE FROM WEEDS

- Weeds compete with the rice plants for the added nitrogen fertilizer.
- Weed before applying nitrogen.
- Weed growth increases with fertilizer application.
- The more vigorous the weed growth the greater will be the competition.

WHY MORE NITROGEN FERTILIZER IS APPLIED DURING THE DRY SEASON

119 Higher grain yields from nitrogen application
120 Less danger of shading
121 Less danger of lodging
122 Increases the low tiller number

HIGHER GRAIN YIELDS FROM NITROGEN APPLICATION

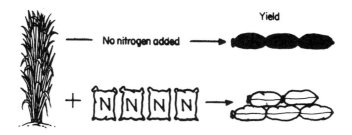

Rainy season

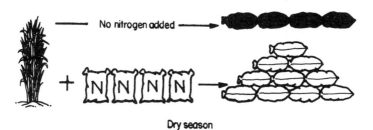

Dry season

- There is a greater response to nitrogen fertilizer during the dry season than during the rainy season.
- Sunlight, which is necessary for the manufacture of food, is more abundant during the dry season.

LESS DANGER OF SHADING

300 units

500 units

Rainy season

Dry season

- Less leafiness — shorter and more erect leaves during the dry season.
- Less danger of shading since there is sufficient sunlight. The amount of light is higher and leaf arrangement for catching the sunlight is better during the dry season.
- If shading occurs yields are reduced.

LESS DANGER OF LODGING

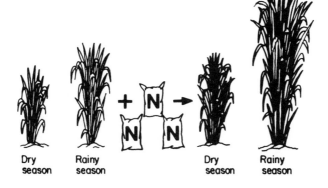

| Dry season | Rainy season | | Dry season | Rainy season |

- Plants do not grow very tall during the dry season compared with the wet season thus lodging is less likely even with higher rates of nitrogen fertilizer.

INCREASES THE LOW TILLER NUMBER

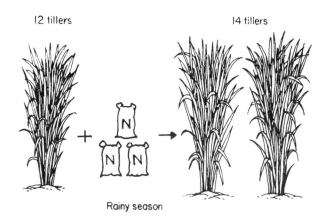

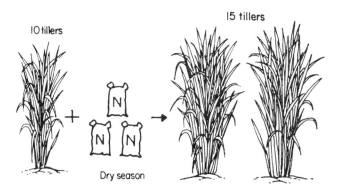

- Nitrogen increases the number of tillers.
- Rice generally produces fewer tillers during the dry than in the rainy season.
- One reason for using closer spacing during the dry season is that fewer tillers are formed during this season.
- The additional tillers produced as a result of nitrogen fertilization are mostly productive since there is less shading during the dry season.

CARBOHYDRATES PRODUCTION

125 The food factory
126 The food factory
127 Factors affecting carbohydrate production — amount of green color
128 Factors affecting carbohydrate production — amount of green color
129 Factors affecting carbohydrate production — amount of light
130 Factors affecting carbohydrate production — amount of light
131 Factors affecting carbohydrate production — amount of water in the leaf
132 Factors affecting carbohydrate production — amount of air

THE FOOD FACTORY

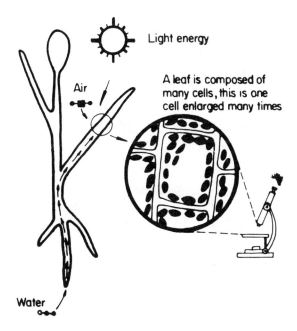

- Carbohydrate is manufactured in the green leaves.
- Water from the soil and carbon dioxide from air are the main materials in the manufacture of carbohydrates.
- Water is absorbed by the roots from the soil. Air enters the plants through the leaf pores on the surface of the leaves.

THE FOOD FACTORY

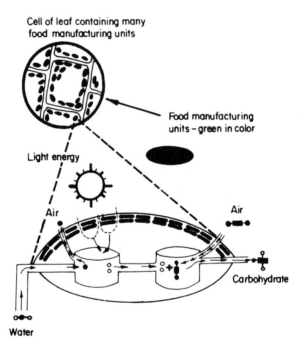

- Light energy is used to break down water which combines with air (carbon dioxide) to produce food.
- The green coloring (chlorophyll) of the leaves collect the light energy.

FACTORS AFFECTING CARBOHYDRATE PRODUCTION— AMOUNT OF GREEN COLOR

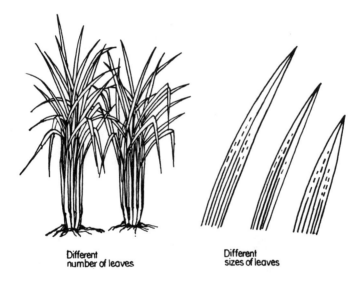

Different number of leaves

Different sizes of leaves

- Total amount of green color (chlorophyll) per plant increases with increase in number and size of the leaves.

FACTORS AFFECTING CARBOHYDRATE PRODUCTION — AMOUNT OF GREEN COLOR

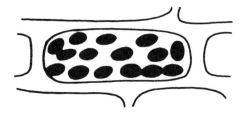

Many manufacturing units

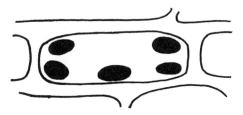

Only few manufacturing units

- Total amount of green color (chlorophyll) per plant increases with increase in thickness of leaves and number of manufacturing units inside the leaves.

FACTORS AFFECTING CARBOHYDRATE PRODUCTION AMOUNT OF LIGHT

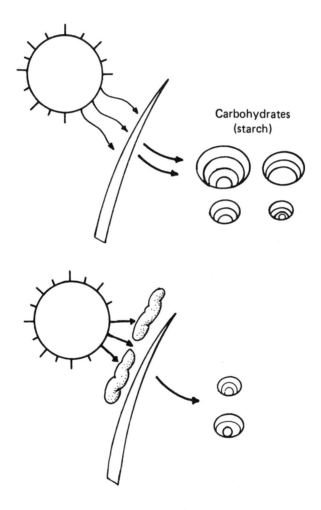

- The brighter the light, the more light energy, resulting in more carbohydrates produced.

FACTORS AFFECTING CARBOHYDRATE PRODUCTION — AMOUNT OF LIGHT

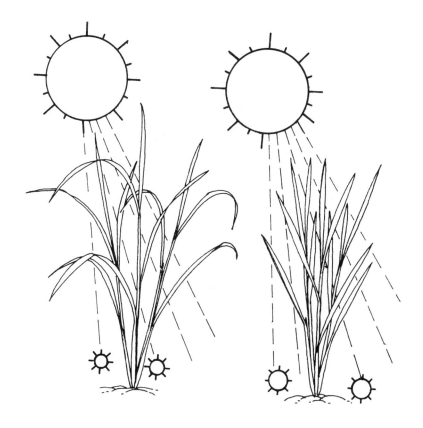

- There is more light available to plants with erect leaves, thus more carbohydrate is manufactured.

FACTORS AFFECTING CARBOHYDRATE PRODUCTION — AMOUNT OF WATER IN THE LEAF

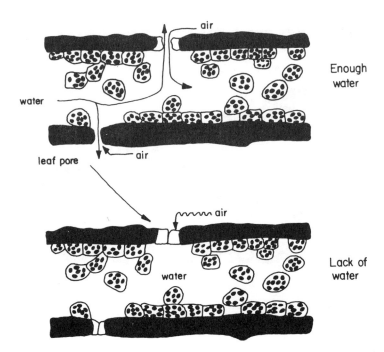

- Loss of water content results in closure of pores.
- Water is an important part of a carbohydrate unit.
- Lack of water leads to decreased rate of food manufacture when pores close and air cannot enter.

FACTORS AFFECTING CARBOHYDRATE PRODUCTION — AMOUNT OF AIR

AMOUNT OF AIR

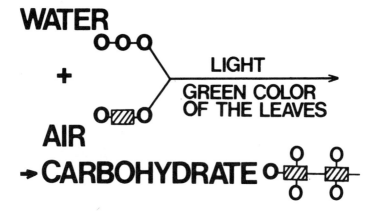

- The plant uses carbon dioxide from the air for food manufacture.
- Carbon dioxide is plentiful and rarely the cause of a decrease in food manufacturing.
- Thus it can be seen that water, air, light, and green color are needed for food manufacture. If any one of these is lacking then food manufacturing is slowed down even if the others are present in abundance.

WATER

135 Major component of the plant
136 Raw material for food manufacturing
137 Carries the food
138 Cools the plant
139 Stiffens the plant

MAJOR COMPONENT OF THE PLANT

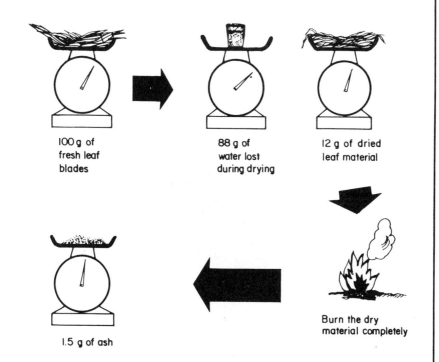

RAW MATERIAL FOR FOOD MANUFACTURING

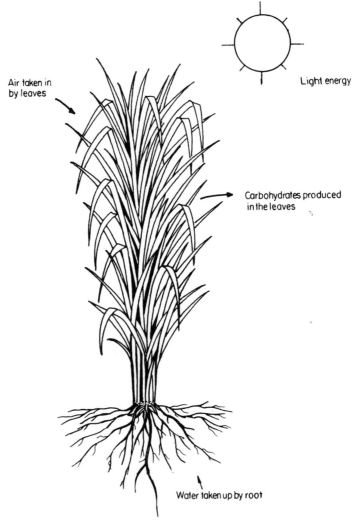

- Lack of water decreases the amount of food manufacturing.
- Water, air, and light are needed in food manufacturing, water is usually the limiting factor.

CARRIES THE FOOD

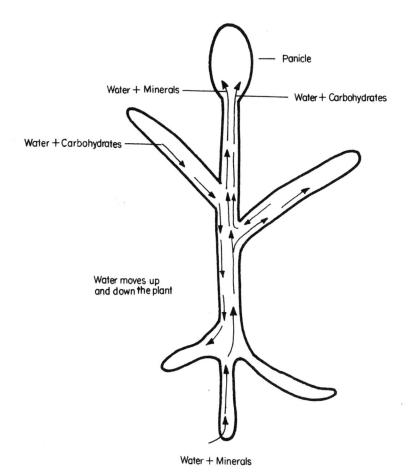

- Water carries the carbohydrates and mineral nutrients to the different plant parts.
- One hectare of rice plants uses at least 8 million liters (400,000 big kerosene cans) of water during its life.

COOLS THE PLANT

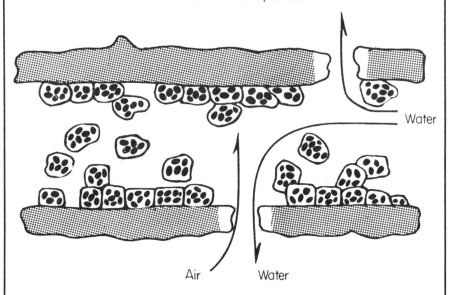

Water – cools the leaf as it evaporates

Cross section of a leaf showing the pores where water evaporates

- Water can cool the leaves similar to perspiration cooling our body.
- Without water in the leaves the pores close: water cannot pass out and air cannot enter. Growth is greatly retarded.
- If the temperature is too high and water does not evaporate, the leaves dry up.
- Most of the water taken up by the rice plant is lost through evaporation.

STIFFENS THE PLANT

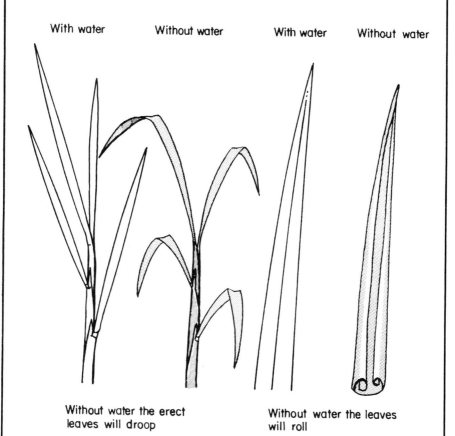

- Water helps in making the leaves erect and fully expanded.
- Water in the plant is like the air in the tires of a car.

YIELD COMPONENTS

- **143** Stages of growth during which yield components are determined
- **144** Stages of growth during which yield components are determined — leaf development and tillering
- **145** Stages of growth during which yield components are determined — panicle formation
- **146** Stages of growth during which yield components are determined — flowering
- **147** Stages of growth during which yield components are determined — ripening
- **148** Variations in yield components
- **149** Importance of yield components
- **150** Importance of yield components
- **151** How to use yield components
- **152** How to use yield components
- **153** How to use yield components
- **154** How to use yield components

STAGES OF GROWTH DURING WHICH YIELD COMPONENTS ARE DETERMINED

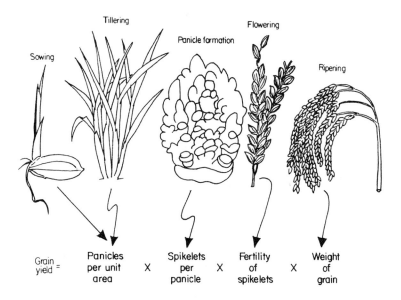

- Every stage of growth contributes to grain yields, good management at all stages is necessary.
- Environmental factors affect all these stages.

STAGES OF GROWTH DURING WHICH YIELD COMPONENTS ARE DETERMINED – LEAF DEVELOPMENT AND TILLERING

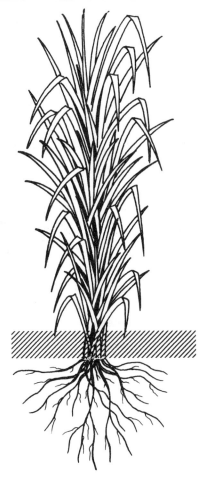

- The number of tillers formed, which determines the number of panicles, is the most important factor in high grain yields.
- Sufficient leaves are necessary to insure large number of spikelets and to fill the spikelets.

STAGES OF GROWTH DURING WHICH YIELD COMPONENTS ARE DETERMINED – PANICLE FORMATION

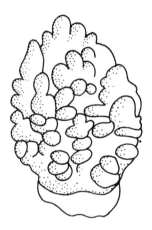

- The number of spikelets per panicle is determined at this stage.
- Very low temperatures and low available light energy during this stage will increase the number of aborted spikelets.

STAGES OF GROWTH DURING WHICH YIELD COMPONENTS ARE DETERMINED – FLOWERING

- Transfer of the male cell to the egg in the ovary occurs at flowering.
- Successful transfer will determine the development of the spikelet into a grain, the result of accumulation of carbohydrate and formation of the embryo.

STAGES OF GROWTH DURING WHICH YIELD COMPONENTS ARE DETERMINED – RIPENING

- The weight of a grain is determined at this stage. It is least affected by environmental factors.
- Poor tillering or low tiller number per unit area cannot be compensated by increasing the weight per grain or increasing fertility of the spikelets since both components do not vary much.

VARIATIONS IN YIELD COMPONENTS

Few but large panicles
panicle weight type

Many but small panicles
panicle number type

- Increase in grain yield of panicle *number* types is usually the result of an increase in number of panicles.
- Increase in grain yield of panicle *weight* types is usually the result of an increase in the weight per panicle.
- Most modern high yielding varieties are panicle number types while the traditional varieties are panicle weight types.

IMPORTANCE OF YIELD COMPONENTS

- Detailed study of the different factors contributing to grain yields can reveal why the yields are high or low.
- Target yield = 80 cavans/hectare or 4,000 kilograms/hectare or 400 grams/square meter

  1 cavan = 50 kilograms of palay

- Characteristics of the variety you are using:
  - number of panicles per hill = 14
  - spikelets per panicle = 100
  - percentage of filled spikelets = 83.3%
  - weight of a single grain = 0.025 grams

IMPORTANCE OF YIELD COMPONENTS

- To find out the number of panicles you need per hill:

$$\text{Yield} = \begin{array}{c}\text{panicles}\\\text{per square}\\\text{meter}\end{array} \times \begin{array}{c}\text{spikelets}\\\text{per}\\\text{panicle}\end{array} \times \begin{array}{c}\text{percent}\\\text{filled}\\\text{spikelets}\end{array} \times \begin{array}{c}\text{weight}\\\text{of a single}\\\text{grain}\end{array}$$

$$400 \text{ g} = (\text{panicles/sq m}) \times (100) \times \frac{83.3}{100} \times (0.025)$$

$$\text{Panicles/sq m} = \frac{400}{100 \times 0.833 \times 0.025}$$

$$= 192$$

- If the spacing you used was 25 × 25 cm or 16 hills/sq m

$$\frac{192 \text{ panicles/sq m}}{16 \text{ hills/sq m}} = 12 \text{ panicles/hill}$$

- The variety you are using can produce more than 12 panicles per hill at 25 × 25 cm spacing. So, target yield could be obtained.

- If yield did not meet target:
  If the yield actually obtained was below 400 grams per square meter although you were using the correct variety and spacing, something was wrong with your crop. A detailed study of the yield components may reveal what was possibly wrong during the growth of the plants.

HOW TO USE YIELD COMPONENTS

THE PROBLEM:

Expected

Actual

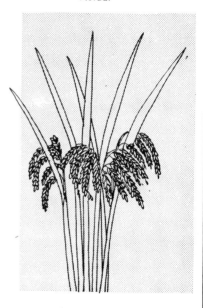

14 panicles

6 panicles

- What was wrong: probably a defect in the soil or in application of fertilizer. It also could have been lack of water or damage by pests and diseases during earlier growth.

HOW TO USE YIELD COMPONENTS

The problem:

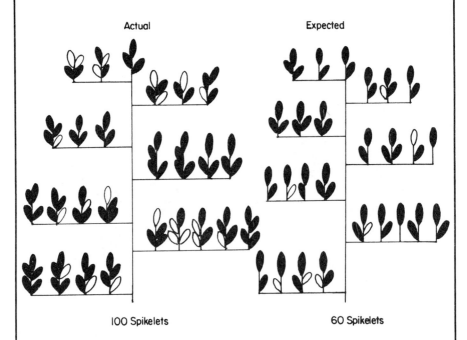

- What was wrong: the problem occurred a little before, during, or after the formation of the spikelet (26 to 16 days before flowering). It possibly resulted from lack of sunlight, lack of food, or insect damage to the leaves.

HOW TO USE YIELD COMPONENTS

The problem:

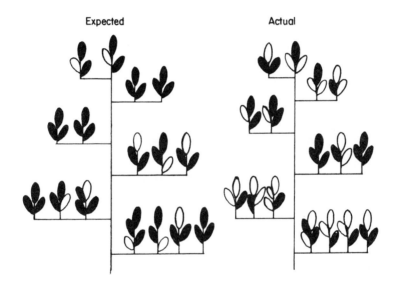

Expected — 80% Filled spikelets

Actual — 50% Filled spikelets

- What was wrong: maybe the temperature was too low (20°C) or too high (above 35°C), the plants lodged, or suffered from lack of water at flowering time. Maybe too much nitrogen was applied.

HOW TO USE YIELD COMPONENTS

Expected

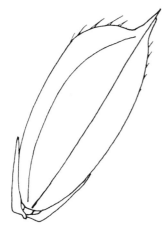

25 grams per 1000 grains

Actual

20 grams per 1000 grains

- What was wrong: conditions after flowering were unfavorable such as not enough food, not enough leaves to manufacture the food, or cloudy weather.

PLANT TYPE OF A LOWLAND RICE VARIETY WITH HIGH GRAIN YIELD POTENTIAL

157 Short stature
158 Non-lodging
159 Good distribution of light
160 Erect leaves
161 Flag leaf higher than the panicle
162 Short leaves
163 Good tillering
164 Erect tillers
165 Ideal tiller

SHORT STATURE

- Reduction in plant height is the most important factor in increasing the grain yield potential of rice.
- Reduction in plant height increases resistance to lodging.

NON-LODGING

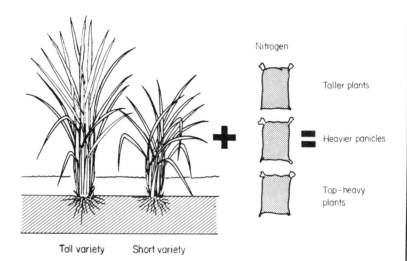

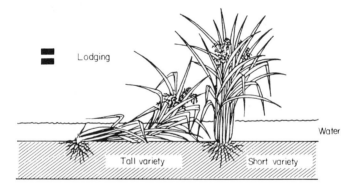

- Plant height increases with nitrogen application and lodging becomes a problem.
- Many leaves on the lodged plants decay since they are soaked in water and do not receive sufficient light.
- Short, stiff stem prevents lodging.

GOOD DISTRIBUTION OF LIGHT

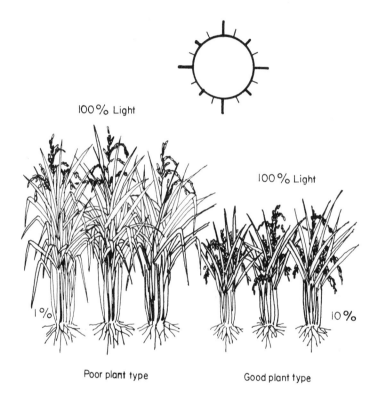

- Tall, leafy plants result in very little light received by the lower leaves.
- Upright tillers and leaves reaching above the panicles contribute to better light distribution, resulting in better food manufacturing and grain yields.

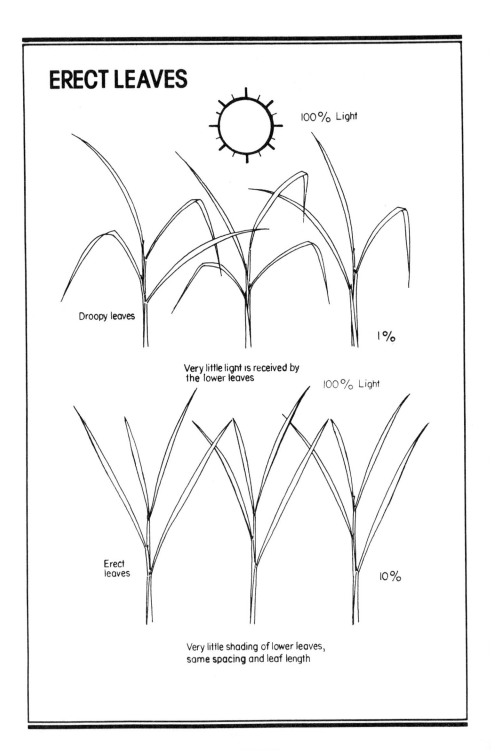

FLAG LEAF HIGHER THAN THE PANICLE

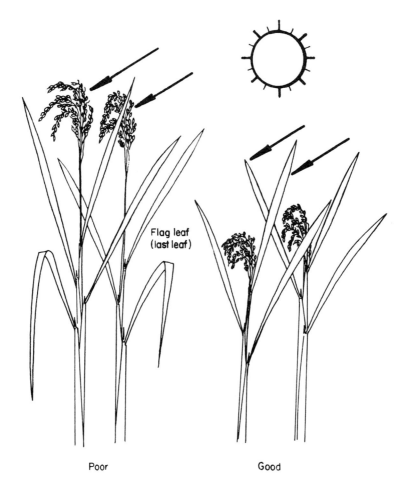

Poor   Good

Erect leaves

- There is less shading of the upper leaves if the panicle does not extend far above the flag leaf.

SHORT LEAVES

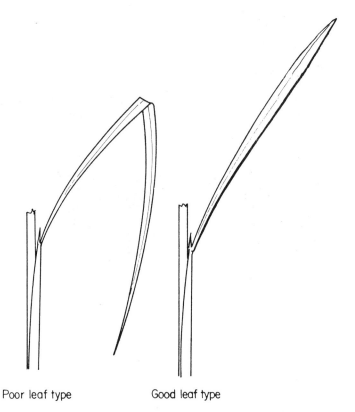

Poor leaf type   Good leaf type

- Shorter leaves are more erect because they have less weight to carry.
- Erect leaves allow more light to reach the lower parts of the plants.

GOOD TILLERING

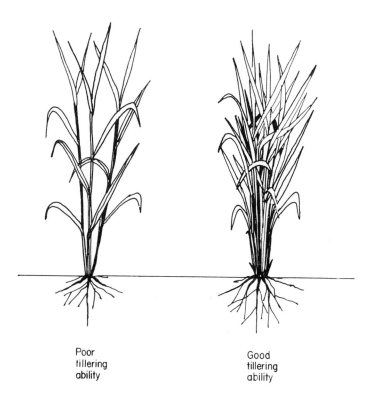

Poor tillering ability

Good tillering ability

- Good tillering ability insures adequate tillers per unit area even if some plants die at early stage of growth.

ERECT TILLERS

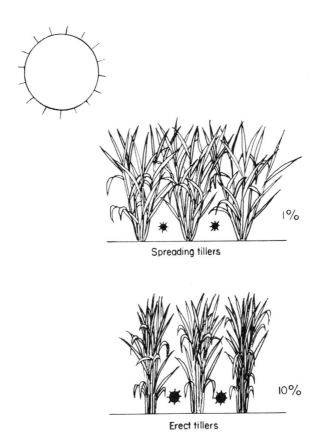

- Upright tillers result in better light distribution.

IDEAL TILLER

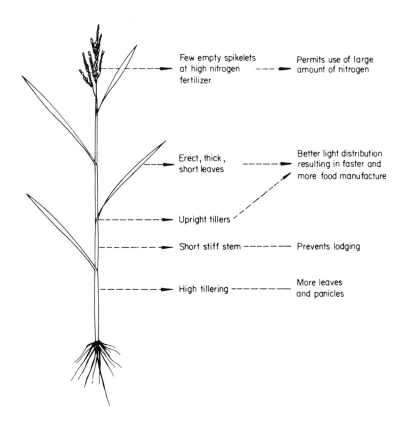

- Main tiller at flowering.

FACTORS AFFECTING LODGING

169 Plant height
170 Method of sowing
171 Type of leaf sheath
172 Stem thickness
173 Wind and rain
174 Light intensity
175 Spacing
176 Amount of fertilizer

PLANT HEIGHT

Not resistant

Resistant

- The taller the plant the greater is the tendency to lodge.
- Avoid using tall varieties during the rainy season.

FACTORS AFFECTING LODGING

METHOD OF SOWING

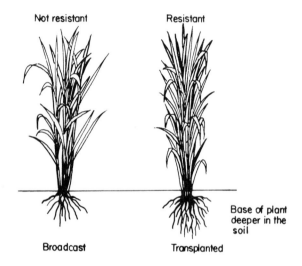

- Transplanted rice is more resistant to lodging because the base of the plant is better anchored.

TYPE OF LEAF SHEATH

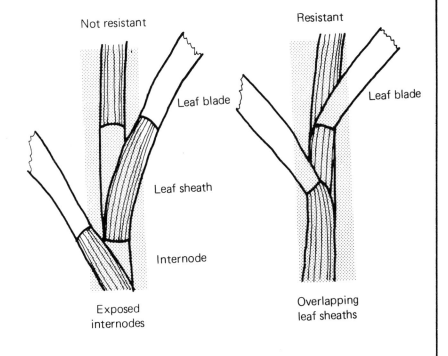

STEM THICKNESS

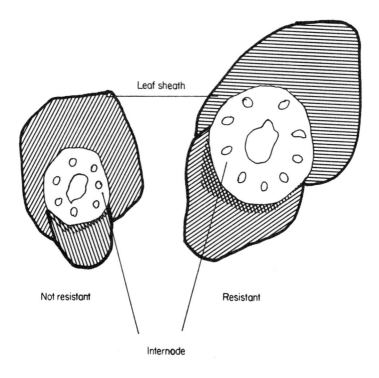

- The thicker the stem and the thicker the internode, the higher the resistance to lodging.

WIND AND RAIN

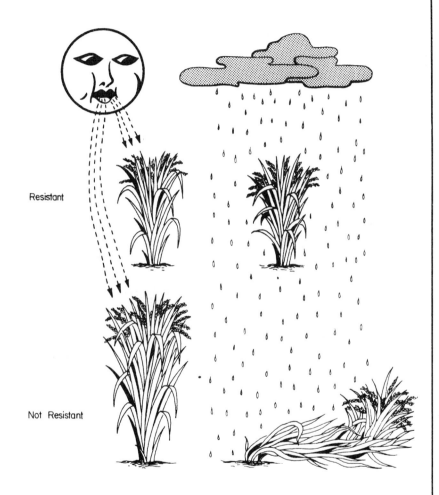

- Wind and rain can lodge a plant. The stronger the wind the more likely the plant will lodge.
- Avoid using tall varieties during the rainy season.

LIGHT INTENSITY

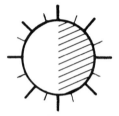

Dry season

Plants shorter

Rainy season

Plants taller

- Cloudy weather results in taller plants, hence greater tendency to lodge.
- Lodging is more common during the rainy season.

SPACING

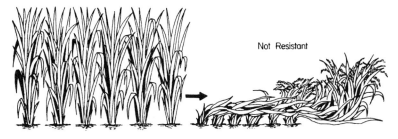

Spacing is too close

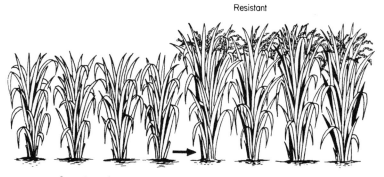

Correct spacing

- Close spacing results in taller plants and weaker stems.

FACTORS AFFECTING LODGING

AMOUNT OF FERTILIZER

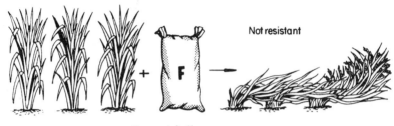

Too much fertilizer

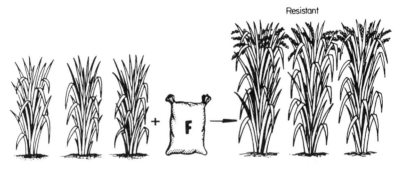

Correct amount of fertilizer

- Fertilizer increases plant height. Tall varieties cannot stand too much fertilizer.

WEEDS

179 Weeds reduce rice yields
180 Weeds compete with rice
181 Weeds decrease the effect of nitrogen fertilizer
182 Weeds — differences between grasses, sedges, and broadleaved weeds
183 Common weeds in rice fields — grass
184 Common weeds in rice fields — sedge
185 Common weeds in rice fields — broadleaf
186 Differences between grasses and rice plants
187 When to weed the rice crop

WEEDS REDUCE RICE YIELDS

Grain yield during the dry season

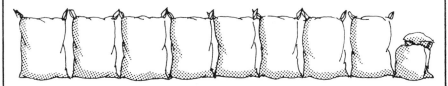

Weeded

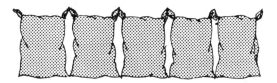

Not weeded

Grain yield during the wet season

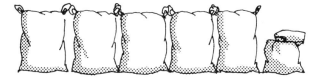

Weeded

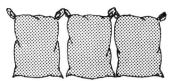

Not weeded

- Weeds reduce grain yields regardless of the season of planting.

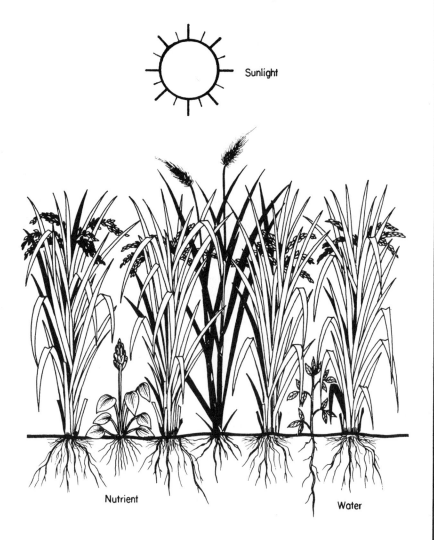

- Weeds compete with rice for sunlight, nutrients, and water.
- If any of these is lacking the others cannot be used effectively even if present in large amounts.
- The competition results in poor rice growth, thus less grain yield.

WEEDS DECREASE THE EFFECT OF NITROGEN FERTILIZER

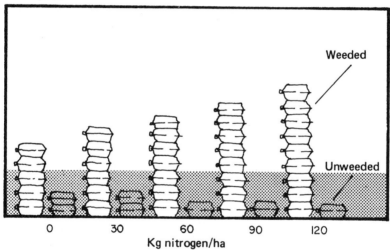

- The higher the amount of nitrogen applied, the less is the grain yield if the crop is not weeded.
- Weeds compete with rice for the applied nitrogen fertilizer.
- Nitrogen application favors the growth of weeds more than the rice crop.
- Nitrogen fertilizers should not be used before weeds are controlled.

WEEDS – DIFFERENCES BETWEEN GRASSES, SEDGES AND BROADLEAVED WEEDS

Type	Grasses	Sedges	Broadleaves
Leaf shape			
Vein arrangement			
Stem cross section			
Examples	Echinochloa crus-galli (Bayakibok)	Cyperus rotundus (Mutha)	Monochoria vaginalis (Gabing uwak)

COMMON WEEDS IN RICE FIELDS — GRASS

Scientific name: *Echinochloa crusgalli*
Common name: Bayakibok

COMMON WEEDS IN RICE FIELDS — SEDGE

Scientific name: *Cyperus iria*
Common name: Sud-sud

COMMON WEEDS IN RICE FIELDS — BROADLEAF

Scientific name: *Monochoria vaginalis*
Common name: Gabing uwak

DIFFERENCES BETWEEN GRASSES AND RICE PLANTS

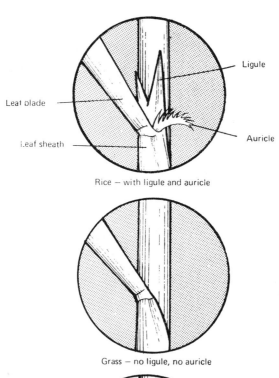

Rice — with ligule and auricle

Grass — no ligule, no auricle

Grass — with ligule, no auricle

WHEN TO WEED THE RICE CROP

Grain yield

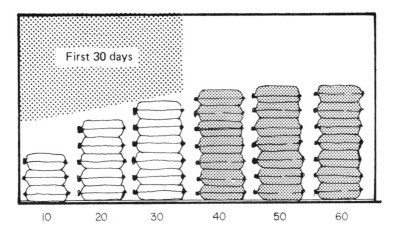

Weed-free days after transplanting

- Weeding in the first 30 days following transplanting of the rice plant is important.
- Grain yield is drastically reduced if rice is not weeded during the early stages of growth.

CONTROL OF WEEDS

191 Weeds can be controlled by hand pulling
192 Weeds can be controlled by mechanical means
193 Weeds can be controlled by proper water management
194 Weeds can be controlled by proper land preparation
195 Weeds can be controlled by crop competition
196 Weeds can be controlled by herbicides

WEEDS CAN BE CONTROLLED BY HAND PULLING

BROADLEAF

Scientific name *Monochoria vaginalis*
Common name Gabing Uwak

- Pulling weeds by hand is a manual method of control.
- Hand pulling is time consuming.

WEEDS CAN BE CONTROLLED BY MECHANICAL MEANS

- Use of rotary weeder is more efficient than hand weeding.
- Straight row planting is necessary if rotary weeder is used.
- Standing water should be drained from the field when the rotary weeder is used.

WEEDS CAN BE CONTROLLED BY PROPER WATER MANAGEMENT

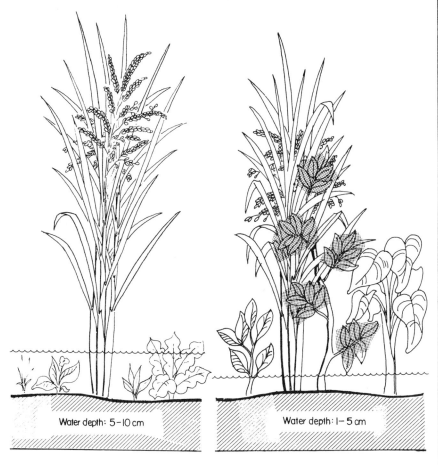

Water depth: 5–10 cm

Growth of weeds greatly reduced

Water depth: 1–5 cm

Growth of weeds slightly reduced

- Most grasses and sedges will be prevented from growing when covered with 5–10 cm water.
- Some broadleaved weeds are not controlled by flooding.
- Many weed seeds do not germinate under water.

WEEDS CAN BE CONTROLLED BY PROPER LAND PREPARATION

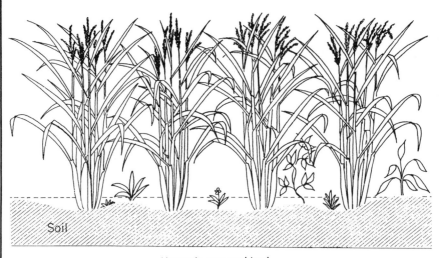

Unevenly prepared land

- Weed growth is encouraged when land is poorly and unevenly prepared and areas are not covered by water.

WEEDS CAN BE CONTROLLED BY CROP COMPETITION

10 x 10 cm spacing

15 x 15 cm spacing

20 x 20 cm spacing

- The closer the plant spacing the fewer the weeds — less light for the weeds to germinate and grow.
- The shorter the weeds the lesser is the weed damage.

WEEDS CAN BE CONTROLLED BY HERBICIDES

Applying powders or liquids in solution

Applying granules

HERBICIDES

199	Types of herbicides based on formulation
200	Types of herbicides based on time of application
201	Types of herbicides based on selectivity
202	Types of herbicides based on types of action
203	Rice injuries from too much herbicide — tillers tend to spread out
204	Rice injuries from too much herbicide — occurrence of brown spots
205	Rice injuries from too much herbicide — formation of onion-like leaves
206	Rice injuries from too much herbicide — dwarfing of the plant
207	Herbicides may kill plants by preventing the manufacture of food
208	Herbicides may kill plants by interfering with the plant system

TYPES OF HERBICIDES BASED ON FORMULATION

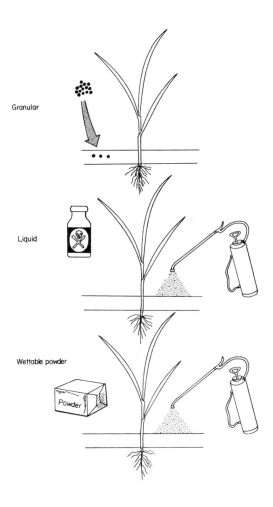

- Commercial herbicides are available in powder, liquid or granular forms.
- Granular forms are broadcast and no special equipment needed for application.

TYPES OF HERBICIDES BASED ON TIME OF APPLICATION

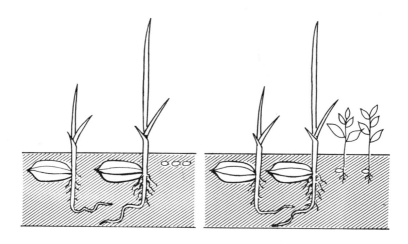

Before the weed seedlings come out.

After the weed seedlings are out.

TYPES OF HERBICIDES BASED ON SELECTIVITY

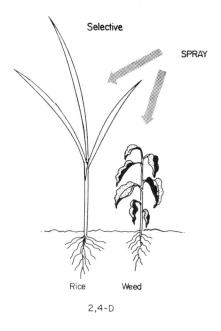

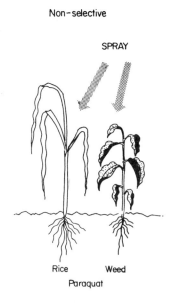

Selective herbicides will kill certain plants only at low concentrations.

Non-selective herbicides will kill all plants.

TYPES OF HERBICIDES BASED ON TYPES OF ACTION

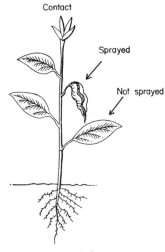

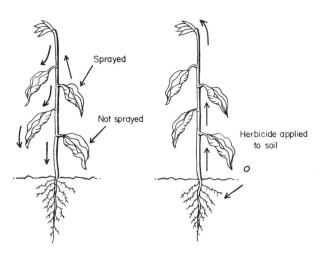

- Contact herbicides will kill only the plant parts that were sprayed.
- Systemic (translocated) herbicides can travel inside the plant and can therefore kill the whole plant.

RICE INJURIES FROM TOO MUCH HERBICIDE – TILLERS TEND TO SPREAD OUT

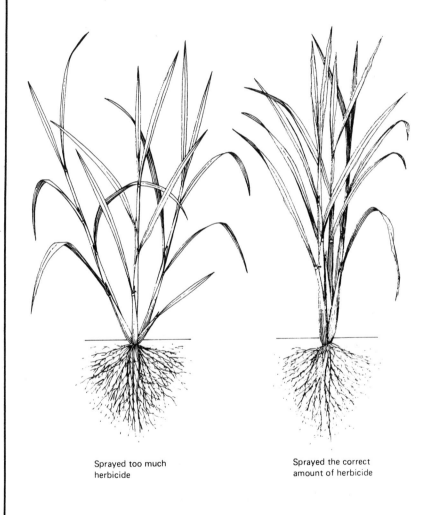

Sprayed too much herbicide

Sprayed the correct amount of herbicide

RICE INJURIES FROM TOO MUCH HERBICIDE — OCCURRENCE OF BROWN SPOTS

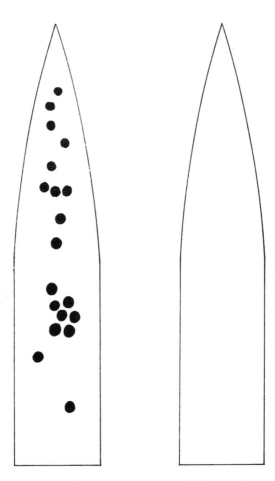

Too much   Right amount

- Herbicide injury may look like a blast or cercospora leaf spot but closer look shows that the spots are discretely round.

RICE INJURIES FROM TOO MUCH HERBICIDE — FORMATION OF ONION-LIKE LEAVES

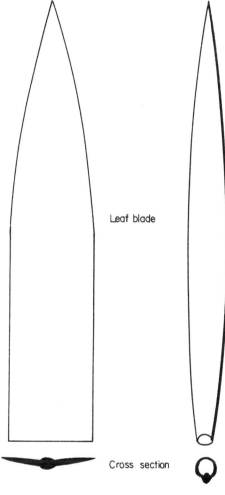

- The new leaves coming out are tube-like or cylindrical if the amount of herbicide applied was too great.

RICE INJURIES FROM TOO MUCH HERBICIDE — DWARFING OF THE PLANT

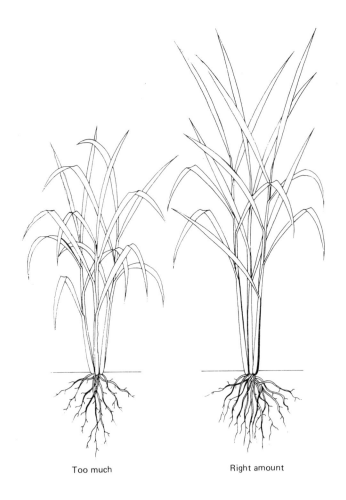

Too much   Right amount

- Be sure to use the correct amount of herbicide. Follow the recommended rate.

HERBICIDES MAY KILL PLANTS BY PREVENTING THE MANUFACTURE OF FOOD

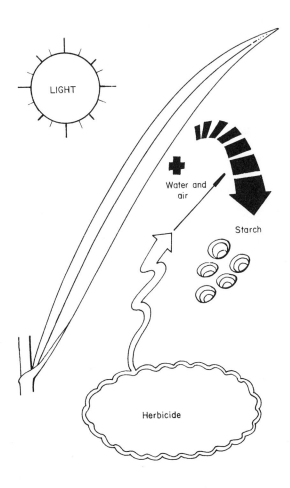

- Herbicides can stop the many different activities going on in the rice plant.
- The manufacture of food involves many steps. A herbicide can block one or more of these steps.

HERBICIDES MAY KILL PLANTS BY INTERFERING WITH THE PLANT SYSTEM

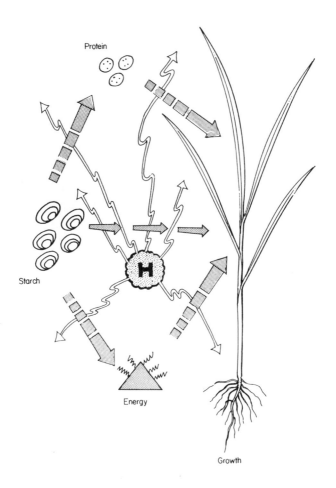

- The manufacture of protein and production of energy leading to plant growth involves hundreds of steps.
- A different protein-compound is responsible for each step. An herbicide might damage any of these protein compounds.
- Any break in the steps may cause death to the plant.

HOW TO JUDGE A RICE CROP AT FLOWERING

211 At flowering a good rice crop should have uniform plant height
212 At flowering a good rice crop should have no lodging
213 Lodging may indicate spacing used was too close
214 Lodging may indicate fertilizer applied was too much
215 Lodging may indicate variety used was too tall
216 At flowering a good rice crop should have white to brown roots
217 At flowering a good rice crop should have green and undamaged leaves
218 At flowering a good rice crop should have at least 3 to 4 leaves per tiller
219 At flowering a good rice crop should have the correct plant density
220 At flowering a good rice crop should have 250 to 350 panicles per square meter

AT FLOWERING
A GOOD RICE CROP SHOULD HAVE
UNIFORM PLANT HEIGHT

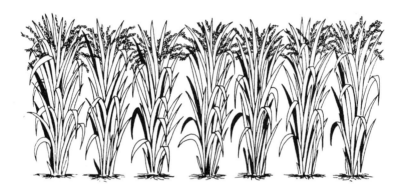

- Irregular plant height can mean many things:
  - plants lacked water or suffered from stem borer attack or virus diseases.
  - uneven land preparation.
  - uneven fertilization.
  - the seeds used were not pure.

AT FLOWERING A GOOD RICE CROP SHOULD HAVE NO LODGING

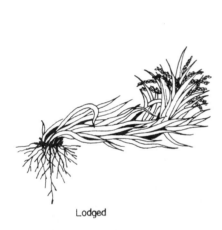
Lodged

Upright

- Lodging may indicate:
  - spacing used was too close
  - too much fertilizer was applied
  - variety used was too tall for that area and for the season of planting.

LODGING MAY INDICATE SPACING USED WAS TOO CLOSE

Spacing too close

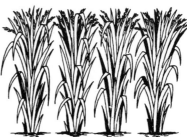

Correct spacing

LODGING MAY INDICATE FERTILIZER APPLIED WAS TOO MUCH

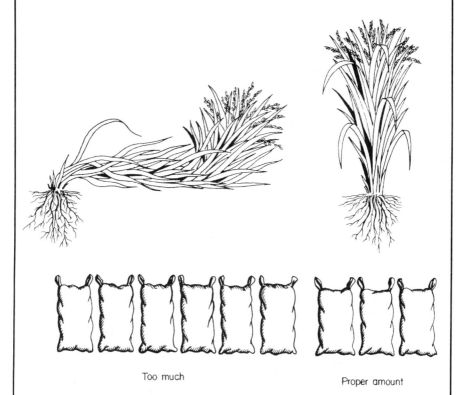

Too much   Proper amount

- Too much fertilizer causes plant to grow too tall, then it lodges.

LODGING MAY INDICATE VARIETY USED WAS TOO TALL

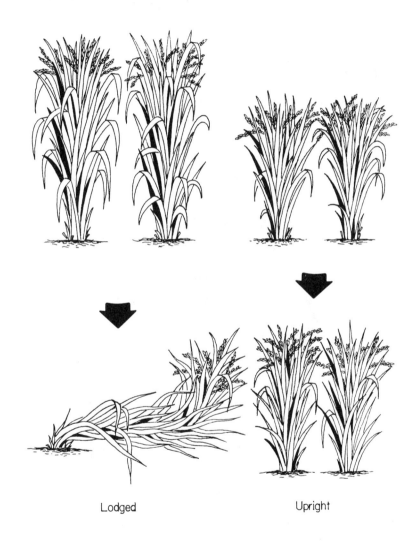

Lodged   Upright

AT FLOWERING
A GOOD RICE CROP SHOULD HAVE
WHITE TO BROWN ROOTS

- Black roots and foul smell indicate something is wrong with the soil:
    - lack of drainage
    - lack of air
    - iron toxicity
    - high organic acids

AT FLOWERING A GOOD RICE CROP SHOULD HAVE GREEN AND UNDAMAGED LEAVES

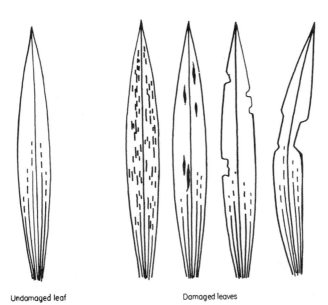

Undamaged leaf   Damaged leaves

- Green undamaged leaves indicate no deficiency or toxicity in the soil, and no attacks by pests or diseases.
- Yellow leaves may indicate nitrogen deficiency or virus diseases.

AT FLOWERING A GOOD RICE CROP SHOULD HAVE AT LEAST 3 TO 4 LEAVES PER TILLER

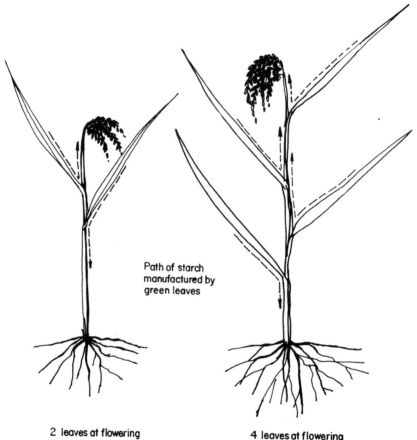

Path of starch manufactured by green leaves

2 leaves at flowering

4 leaves at flowering

- A tiller needs 3 to 4 leaves:
  - to better provide the roots and other plant parts with sufficient food.
  - to better fill up the spikelets with starch manufactured in the leaves.
- If a tiller has only 2 leaves, suspect some deficiency in the soil or water stress at an earlier stage of growth.

AT FLOWERING A GOOD RICE CROP SHOULD HAVE THE CORRECT PLANT DENSITY

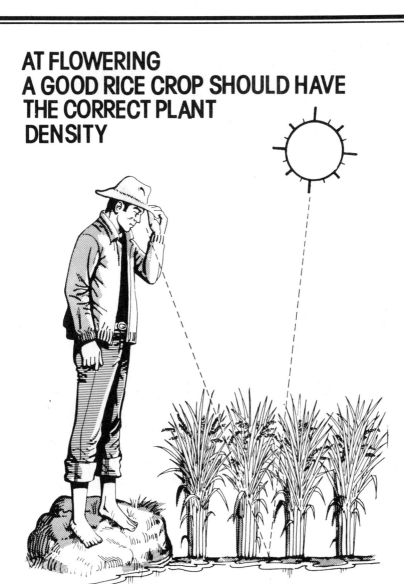

- The correct number of plants per unit area can be checked by standing on the levee. If one can barely see the water or sparkle of the sun rays, then the plant density is right.
- If one cannot see the water, probably the spacing was too close, too much fertilizer was applied, or the variety was too tall.

AT FLOWERING A GOOD RICE CROP SHOULD HAVE 250 TO 350 PANICLES PER SQUARE METER

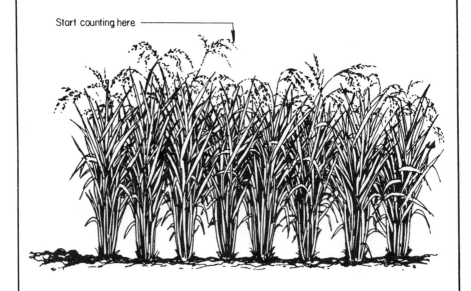

- Count the number of panicles per hill (clump) in at least 3 hills inside the field. Do not count the first 3 rows from the levee.
- Get the spacing used.

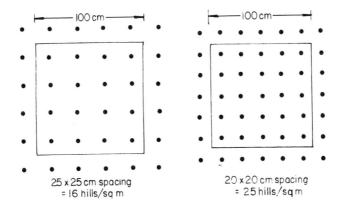

25 x 25 cm spacing
= 16 hills/sq m

20 x 20 cm spacing
= 25 hills/sq m

If the distance between hills is: 25 X 25 cm

Area per hill = 25 X 25 = 625 sq m = 0.0625 sq m

Number of hills per sq m = $\dfrac{1 \text{ sq m}}{\text{area per hill}}$

$= \dfrac{1}{0.0625}$

$= 16$

- To get the number of panicles per square meter

  Assuming: 17 panicles per hill
  16 hills per sq m

Number of panicles per sq m = Number of panicles per hill X Number of hills per sq m

$= 17 \times 16$

$= 272$

- If number of panicles per square meter is less than 250 something is wrong with the method of farming, the rice variety or the soil. Check spacing and fertilizer application.